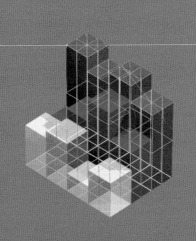

digital diagrams

TREVOR BOUNFORD

SERIES CONSULTANT
ALASTAIR CAMPBELL

CASSELL&CO

To Sheila, Eleanor, Hannah and Felix

Trademarks and trademark names are used throughout this book to describe and inform you about various proprietary products. Such mention is intended to benefit the owner of the trademark and is not intended to infringe copyright nor to imply any claim to the mark other than that made by its owner.

The information contained in this book is given without warranty and, while every precaution has been taken in compiling the book, neither the author nor publisher assume any responsibility or liability whatsoever to any person or entity with respect to any errors which may exist in the book, nor for any loss of data which may occur as the result of such errors or for the efficacy or performance of any product or process described in the book.

A CIP catalogue record for this book is available from the British Library

ISBN 0304 354074

This book was conceived, designed and produced by Alastair Campbell and The Ivy Press Limited The Old Candlemakers West Street, Lewes East Sussex BN7 2NZ

Editorial Director:
Sophie Collins
Art Director:
Peter Bridgewater
Managing Editor:
Andrew Kirk
Editor:
Peter Leek
Designers:
Trevor Bounford and Alastair Campbell
Illustrations:
Denise Goodey, Nicholas Rowland, Helen Humphries and the late Bob Chapman. Particular thanks to Denise Goodey for frequent heroic efforts and to Bruce Robertson for his brutal encouragement. Thanks also to Sarah Lubikowski and Stephen Coulson for help in the early stages.

Originated and printed by Hong Kong Graphic, Hong Kong

Cassell & Co
Wellington House
125 Strand
London WC2R 0BB

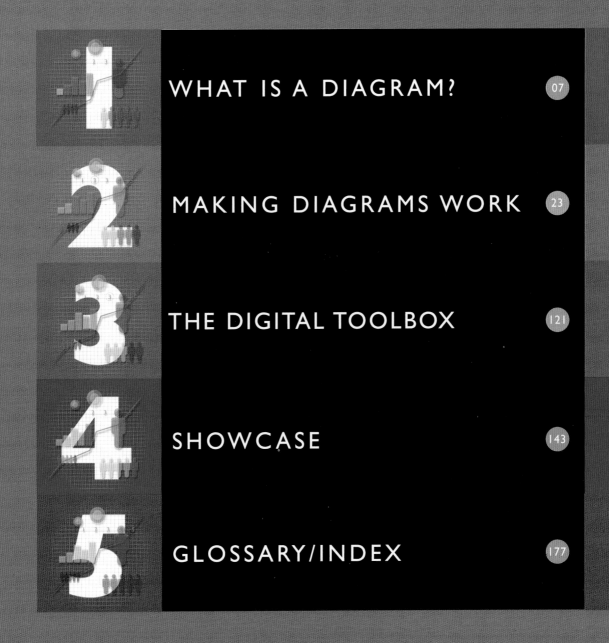

INTRODUCTION

We seem to live in an information driven world. No political statement, company policy or marketing strategy is considered valid without a suitably bulging packet of statistical analysis to support it. Indeed, society seems to have an insatiable appetite for information. However, information in its raw form is largely indigestible for any but the specialist. According to Seurat, 'nature needs a little cooking'; and so it is with data. And the same digital technology that has enabled the glut of data can also be harnessed to distribute it in an appetizing and (informatively) nutritious way. However, although any number of software packages now seem capable of producing data in some graphic format, the expertise that is needed to turn a stew of facts into an appealing display of digestible information is not so readily available. Over the years there have been master chefs in the art of information design. William Playfair and Willard Brinton were great innovators in their time. More recently we have Otto Neurath's Isotype Institute and Bruce Robertson's Diagram Group.

This book is aimed at the non-specialist, the general cook who needs to prepare some specialist dishes. It is, in a way, a recipe book, a book of practical menus. It is a guidebook, a reference book for practical use aimed at pragmatic users. It is not an academic analysis of the subject, nor is it a step-by-step manual for the novice. Apart from the historical examples in Section 1 and one or two items in the Illustrative Diagrams section, all the images have been created digitally. I have assumed that readers will have a reasonable level of competence in computer usage. The book is crammed with as much guidance, as many words and visual examples, as is sensible in a book of this size, and I believe that it will serve to inform and inspire most users. *Bon appetit!*

Trevor Bounford
Cambridge

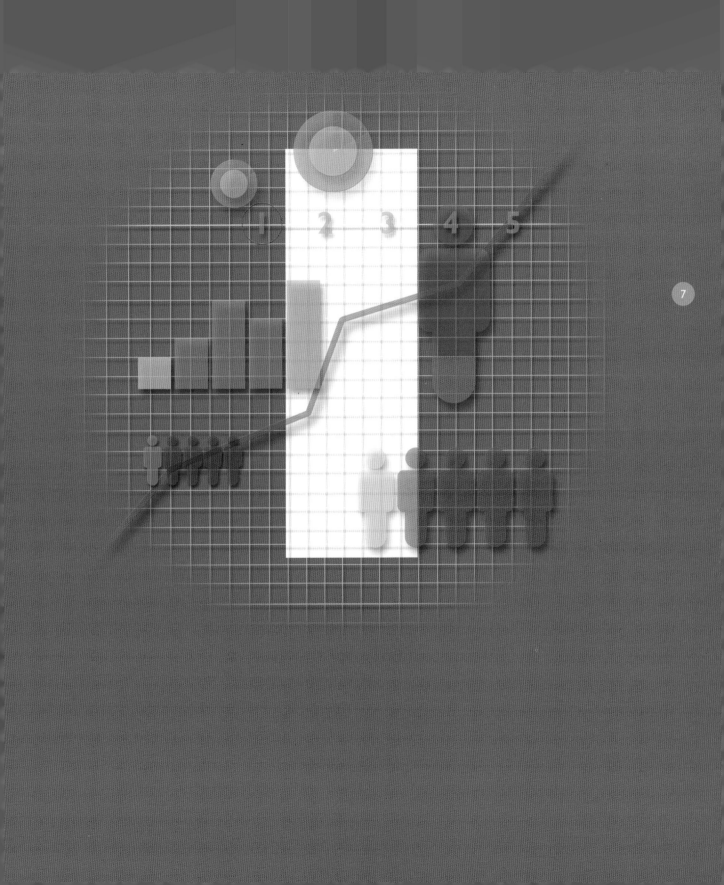

WHAT IS A DIAGRAM?

Deciding what is and what is not a diagram may seem an easy, and perhaps pointless, exercise. However, in order to compile a book about information design, it is important to determine where the vague, and subjective, border might lie. Bar and pie charts, line graphs and other statistical charts are obviously included, as are organizational charts. I also count symbols and icons as valid candidates; there is similarly no problem with timelines; and tables qualify, despite often being largely typographic designs. Most of the borderline cases are, in fact, relational and illustrative diagrams.

When they are cross-sections, exploded or ghosted drawings, or labelled illustrations, illustration-based diagrams are clearly identifiable as diagrams. But what of pure illustration – unlabelled, representational art? If the art is designed in such a way that it, literally, illustrates one or more points of information, or if it is part of a sequence that sets out to do that, I count it as an illustrative diagram. If it merely records a scene, then for the purposes of this book it's an illustration.

For present purposes, I also have a cut-off point for relational diagrams. I've excluded cartography in its conventional form, as found in general atlases, road maps, maritime charts, star charts, and so on. But I do include maps, plans and charts that are designed to address specific information-dissemination tasks. Also, if a map or chart has unique design qualities, I count it as information design, distinguishing it from the kind of standard cartography that is merely the application of an established language. However, I have tried to indicate the importance of the standard cartographic conventions.

In this book, diagrams have been grouped under five headings – illustrative diagrams, statistical diagrams, relational diagrams, organizational diagrams and time charts. But these categories overlap, and they are often combined in individual charts.

GEBURTEN UND STERBEFÄLLE IN WIEN

1912-15

1916-19

1920-23

1924-27

Jedes Kind = 20.000 Lebendgeburten Jeder Sarg = 20.000 Sterbefälle

Above A bar chart composed of icons. Designed in the late 1920s by the Isotype Institute. Although composed of illustrative items, this is clearly a statistical diagram.

Do not walk on the line unless you must. Where there is an authorised route – use it. If you have to walk on the line, keep clear of the tracks whenever possible, and face the direction from which trains approach. Give careful thought to this and make sure that you consider the various possibilities.

WRONG   *Walking in the direction of traffic*

RIGHT   *In the cess and well clear*

3

Left Pages from a safety training manual. More than mere pictures. The colour highlighting makes the point in these illustrative diagrams.

9

RIGHT   *Standing clear*

Wherever you are on the track, it will always pay you to look around and to weigh up your position. Remember that diesel and electric trains and locomotives differ from steam locomotives in several important ways –

they approach quietly at all speeds
there is no smoke or steam
there is little vibration of the rails
they accelerate quickly

Stand well away from any passing train – otherwise you may be hurt by a loose sheet end, or by a sheet-tie, or by a piece of coal or some other object falling from the train.

Keep well clear of trains passing over detonators, or you might be hit by flying pieces of metal.

When on the line, don't get so absorbed in talking with your mates that you neglect your safety – and theirs.

WRONG   *The conversation may be necessary*
                         but it is not in the right place

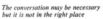

5

ILLUSTRATIVE DIAGRAMS

The term illustrative diagram encompasses a broad spectrum – from simple symbolic icons through to complex cross-sections. Some categories of illustrative diagram (architectural, anatomical, scientific, engineering drawing, and so on) are too specialized to be covered in a book like this. The varieties included here are of the type the information designer is likely to encounter most frequently. Even so, many of them will call for the specialist skills of a trained illustrator, or need access to a comprehensive source of professionally produced clip art that can be digitally manipulated to suit the designer's requirements. Although other types of chart (statistical and organizational diagrams, for example) often include pictorial elements, illustrative diagrams make use of the image itself to describe the situation or events shown. They are generally used to portray physical rather than abstract concepts, and they almost invariably deal with a localized context.

10

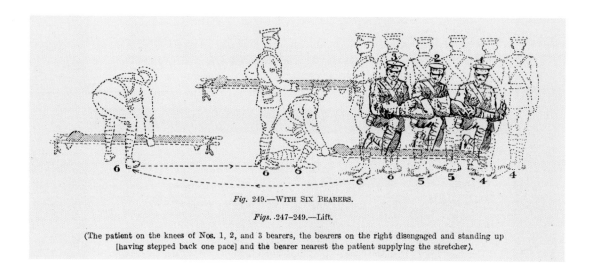

Fig. 249.—WITH SIX BEARERS.

Figs. 247–249.—Lift.

(The patient on the knees of Nos. 1, 2, and 3 bearers, the bearers on the right disengaged and standing up [having stepped back one pace] and the bearer nearest the patient supplying the stretcher).

Above Illustration from the eighth edition of *First Aid to the Sick and Injured*, published in 1913. A highly detailed figure drawing with arrows and labels to convey the instruction.

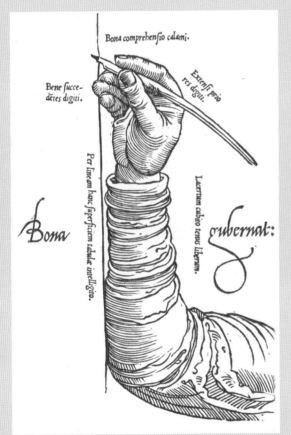

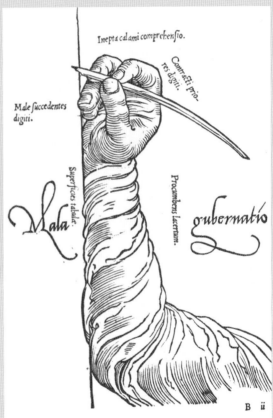

Above Diagrams from a book on italic lettering by Gerard Mercator, published in 1540. The drawings show the right and wrong way to hold the quill for lettering on maps.

TABLES AND GRAPHS

Statistical diagrams are used to enable comparisons of data to be made. In this book, statistical diagrams have been subdivided into the categories of tables, line and scatter graphs, and bar, area and volume charts.

In basic tables the data is set out on a matrix or grid, so that numerical values can be read against category headings. In effect, the table is a set of lists that can be read horizontally or vertically, the user being required to assimilate and assess the values. For convenience, the table section of this book includes non-data tables. These function in much the same way – the difference being that, instead of numerical values, some or all of the table entries are verbal statements or else visual symbols that represent words.

Line graphs make use of a different grid matrix. Each string or list of data is plotted against two intersecting axes. Commonly the horizontal (x) axis represents even divisions of time, and the vertical (y) axis shows values.

TABLE I—REACH

Case	DESCRIPTION	Distance Moved Inches	LEVELED TIME T. M. U.					
			A Std.	A Hand In Motion	A with C D or B	B Hand In Motion	C or D	E
A	Reach to object in fixed location, or to object in other hand or on which other hand rests.	1	1.8	1.3	2.1	1.5	3.6	1.7
		2	3.7	2.8	4.3	2.7	5.9	3.8
		3	5.0	3.8	5.9	3.6	7.3	5.3
		4	6.1	4.9	7.1	4.3	8.4	6.8
B	Reach to single object in location which may vary slightly from cycle to cycle.	5	6.5	5.3	7.8	5.0	9.4	7.4
		6	7.0	5.7	8.6	5.7	10.1	8.0
		7	7.4	6.1	9.3	6.5	10.8	8.7
		8	7.9	6.5	10.1	7.2	11.5	9.3
C	Reach to object in group.	9	8.3	6.9	10.8	7.9	12.2	9.9
		10	8.7	7.3	11.5	8.6	12.9	10.5
		12	9.6	8.1	12.9	10.1	14.2	11.8
		14	10.5	8.9	14.4	11.5	15.6	13.0
D	Reach to very small object or where accurate grasp is required.	16	11.4	9.7	15.8	12.9	17.0	14.2
		18	12.3	10.5	17.2	14.4	18.4	15.5
		20	13.1	11.3	18.6	15.8	19.8	16.7
E	Reach to indefinite location to get hand in position for body balance or next motion or out of way.	22	14.0	12.1	20.1	17.3	21.2	18.0
		24	14.9	12.9	21.5	18.8	22.5	19.2
		26	15.8	13.7	22.9	20.2	23.9	20.4
		28	16.7	14.5	24.4	21.7	25.3	21.7
		30	17.5	15.3	25.8	23.2	26.7	22.9

Left Table from a methods/time measurement system developed in the 1940s. Well-ordered type sets out technical detail so that it may be easily assimilated.

A line connects each of the data points on the graph, and the angle of each line segment indicates the degree of change between the values.

Each line of a line graph must represent variations of the value of an item or a collective group of items. It would be misleading, even nonsensical, to construct a line graph with each data point representing a different category. The data points must be a linear string, so that neighbouring points have a valid proximity.

Scatter graphs (*see page 52*) are similarly based on two axes – but instead of a linear string, the values are for individual items and the points are not connected by lines. It is the interrelationship of the various points that is significant, rather than the progressive comparison of the line graph.

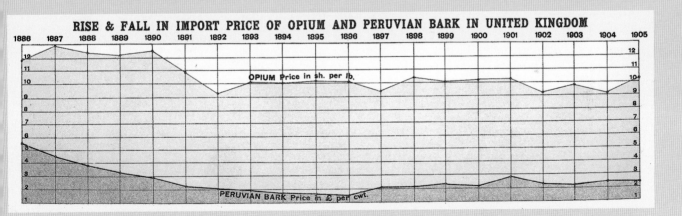

Above Line graph from *Atlas of the World's Commerce*, published in 1907. A straightforward line graph which is more interesting for its content than for its treatment.

BAR, AREA AND VOLUME CHARTS

The three other varieties of statistical diagram to be found in this book are the bar chart, the area chart and the volume chart. These portray the values of items as length (in bar charts), area or volume. In effect, they display values in one, two or three dimensions. Whereas bar charts and area charts are shown in their entirety in a two-dimensional drawing (either in print or on screen), volume charts have to convey the concept of the third dimension within the constraints of two dimensions.

Unlike the line graph, which presents a sequential review of the value of a single item or category, the bar chart has the flexibility to represent a series of different (but comparable) items. For example, a bar chart could show the price of a commodity at defined points over a given period of time. Or it could show the price of the same commodity over the same period but in different locations. Or it could show the prices of a range of commodities in a given place or over a given period,

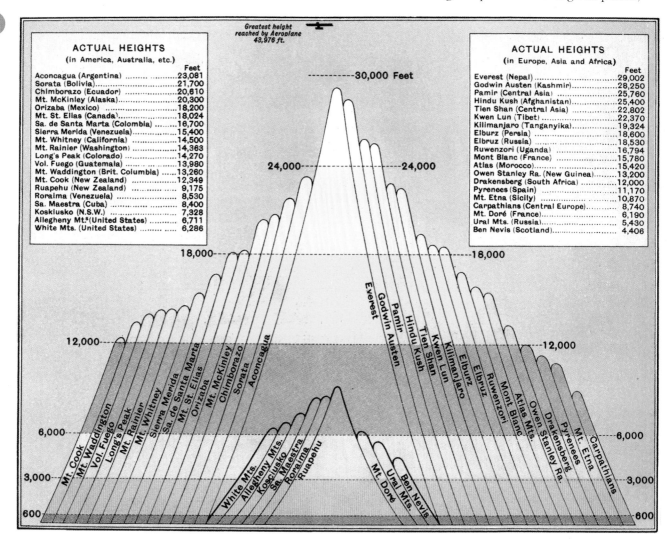

ACTUAL HEIGHTS
(in America, Australia, etc.)

	Feet
Aconcagua (Argentina)	23,081
Sorata (Bolivia)	21,700
Chimborazo (Ecuador)	20,610
Mt. McKinley (Alaska)	20,300
Orizaba (Mexico)	18,200
Mt. St. Elias (Canada)	18,024
Sa. de Santa Marta (Colombia)	16,700
Sierra Merida (Venezuela)	15,400
Mt. Whitney (California)	14,500
Mt. Rainier (Washington)	14,363
Long's Peak (Colorado)	14,270
Vol. Fuego (Guatemala)	13,980
Mt. Waddington (Brit. Columbia)	13,260
Mt. Cook (New Zealand)	12,349
Ruapehu (New Zealand)	9,175
Roraima (Venezuela)	8,530
Sa. Maestra (Cuba)	8,400
Kosciusko (N.S.W.)	7,328
Allegheny Mts. (United States)	6,711
White Mts. (United States)	6,286

ACTUAL HEIGHTS
(in Europe, Asia and Africa)

	Feet
Everest (Nepal)	29,002
Godwin Austen (Kashmir)	28,250
Pamir (Central Asia)	25,760
Hindu Kush (Afghanistan)	25,400
Tien Shan (Central Asia)	22,802
Kwen Lun (Tibet)	22,370
Kilimanjaro (Tanganyika)	19,324
Elburz (Persia)	18,600
Elbruz (Russia)	18,530
Ruwenzori (Uganda)	16,794
Mont Blanc (France)	15,780
Atlas (Morocco)	15,420
Owen Stanley Ra. (New Guinea)	13,200
Drakensberg (South Africa)	12,000
Pyrenees (Spain)	11,170
Mt. Etna (Sicily)	10,870
Carpathians (Central Europe)	8,740
Mt. Doré (France)	6,190
Ural Mts. (Russia)	5,430
Ben Nevis (Scotland)	4,406

Greatest height reached by Aeroplane 43,976 ft.

and so on. It is this flexibility that makes it particularly useful as an information-design tool.

Area charts and volume charts share this flexibility. Their particular advantage is that they can accommodate a greater range of values. A linear value of 100 units is only 10 times the length of a linear value of 10 – but an area of 100 units is the equivalent of 10 units across by 10 high, and a volume measuring 10 across by 10 high by 10 deep represents a value of 1000.

The most common form of area chart is the pie chart (sometimes called a cake chart), which in its most basic form is just a radially divided circle. Volume charts are used much less frequently than area charts, even though modern digital technology facilitates accurate three-dimensional plotting.

15

AIR FLEETS OF THE WORLD *

United States 1,752

France 1,687

Russia 1,500

Japan 1,384

Great Britain 838   Yugoslavia 627   Rumania 599

Czechoslovakia 546   Spain 462   Canada 355

* "First Line" Military Aeroplanes

RELATIONAL DIAGRAMS

There are two kinds of diagrams that illustrate the relationship of non-quantitative items. Relational diagrams show the relative positions of things that are either at fixed locations in the physical world or are planned to be at fixed locations (whereas organizational diagrams illustrate the conceptual interrelationship of physical or non-physical entities regardless of their geographical location).

Signs, plans, maps and charts all provide a graphic indication or representation of location. They enable a user to relate to places, and to relate those places to each other. In one form or another, maps have existed for thousands of years. Made from bark, hide, bamboo, papyrus, or other durable materials, they have enabled humans to travel great distances. Early maps of the world may lack precision in measurement, but in concept they are remarkably accurate. Even the stars have been mapped from early times, as an aid to navigation.

Signs along routes – originally temporary mounds of pebbles, but later permanent objects, painted or carved with names and distances – are relational implements. The modern equivalent, the motorway indicator with a variable digital display, is the current version of a visual tool originating in prehistory.

Digital technology allows more sophisticated and more accurate representation, yet by their very nature relational diagrams are inevitably a compromise. They are a reduction of the physical world to a manageable but recognizable visual form. For the sake of clarity and functional efficiency, not all detail can be included. The reality has to be edited, and it is the information designer's function to make editorial decisions that will result in a plan, map or chart that fulfils its function by aiding, and not hindering, the user.

16

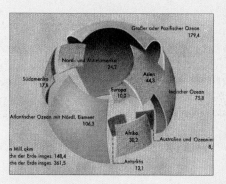

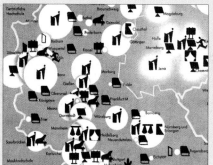

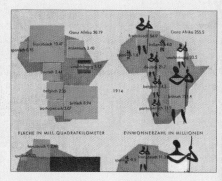

Right and above
Sections of maps from *Grosser Weltatlas der Büchergilde*, published in 1963. A wide variety of highly stylized treatments intended to maintain interest throughout a large volume of diagrams and maps.

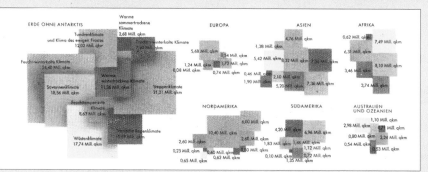

Right Map of the heavens in August, September and October, from Milner's *Gallery of Nature*, published in 1855. An immensely detailed hand-engraving.

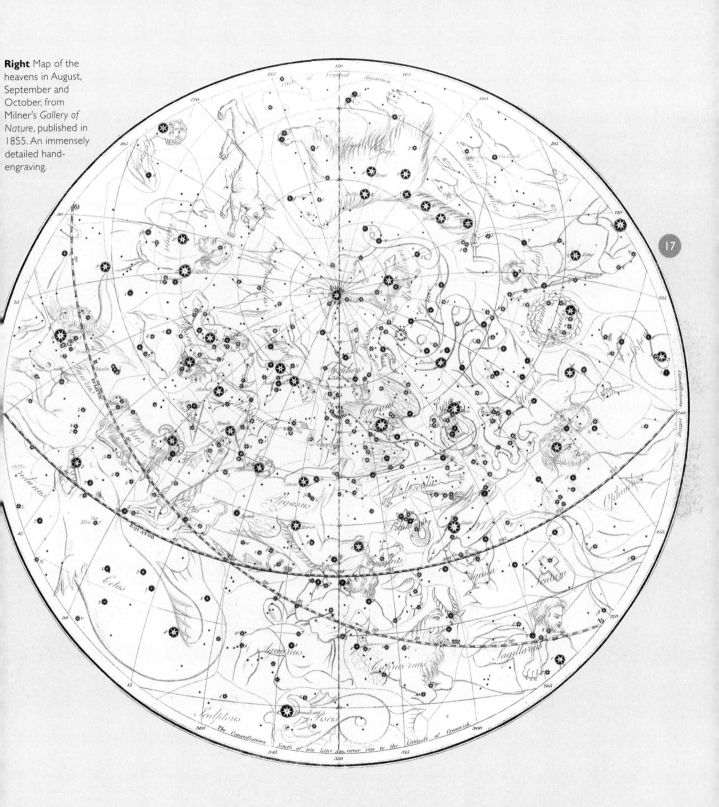

ORGANIZATIONAL DIAGRAMS

We live in a world of complex social structures – local, national and international governments and institutions, multinational commercial enterprises, non-governmental organizations, and so on. Organizational diagrams are needed to explain the complex and often convoluted links within and between these structures.

There are numerous applications for organizational charts. The family tree is a commonly used and relatively simple form. It sets out the lines of descent through two or (usually) more generations. Complex versions may weave together lines from several families. Charts of the same type are frequently used to plot the corporate interminglings of commercial enterprises resulting from mergers and takeovers.

Organizational charts are also employed to explain or summarize cyclical flow and interaction – for example, to describe the natural water cycle, a continuous process with no beginning and (hopefully) no end.

Unlike relational diagrams, which deal with physical items only, organizational diagrams can be used to show the relationship of activities, concepts or abstract items. They can also, in the form of flow charts, be used to describe stages in a process.

Diagnostic charts (a type of organizational chart able to encompass abstract processes – thoughts, feelings, and so on) plot a course of considerations or alternative 'routes', determined by answers to questions embedded within the chart.

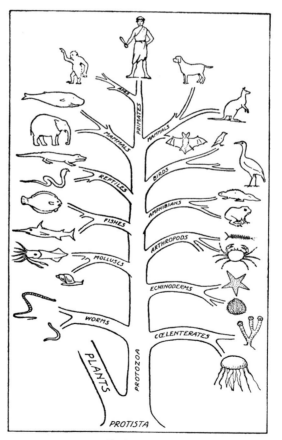

The Tree of Life

Left The Tree of Life, from *The Wonderful Story of the Human Body*, by 'A Well Known Physician'. A simple organization chart reducing a complex structure to a simple visual analogy.

Right Organizational diagram designed by the Isotype Institute in 1947 for *How Do You Do Tovarish?*, published by Adprint Ltd/George Harrap and Co. Ltd. The Isotype Institute set out to convey information with the minimum amount of 'written' language.

Opportunity in the Soviet Union

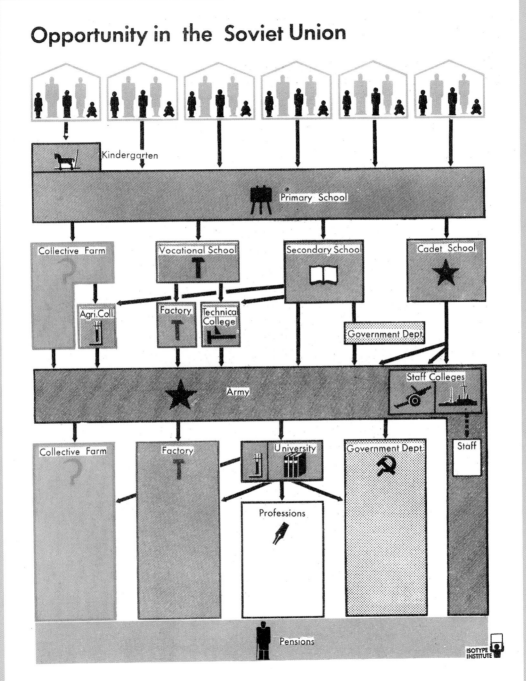

19

11

TIME CHARTS

Time may be portrayed visually in a number of ways. Generally we perceive time as a linear progression. Indeed, the term 'timeline', denoting a linear visual record of events, derives from this perception. But time can also be cyclical, with recurring events following a circular pattern. We also use the table matrix to visualize time, in diaries, calendars and timetables.

All of these visualizations are valid, and each has its own uses and context. It's informative to plot the course of historical events along a linear path, particularly when making comparisons over a period of time – the changing fortunes of different civilizations, for example, or phases of development or cultural activity within a single civilization. Timetables list a series of comparative options, with the time progression shown along both the horizontal and vertical axes. Cyclical events can be shown on a circular grid (seasonal activity following a natural cycle particularly lends itself to this).

20

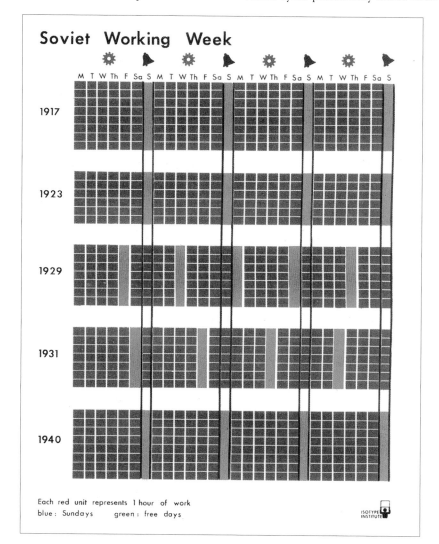

Right Time chart designed by the Isotype Institute, from *How Do You Do Tovarish?*, published in 1947 by Adprint Ltd/ George Harrap and Co. Ltd. Information simplified to a visual pattern that can be readily assimilated.

Right Time chart from *Graphic Methods for Presenting Facts* by Willard C. Brinton, published in 1914 by McGraw-Hill. Again, data reduced to a visual pattern allowing the user to read detail and obtain a clear impression of the implications.

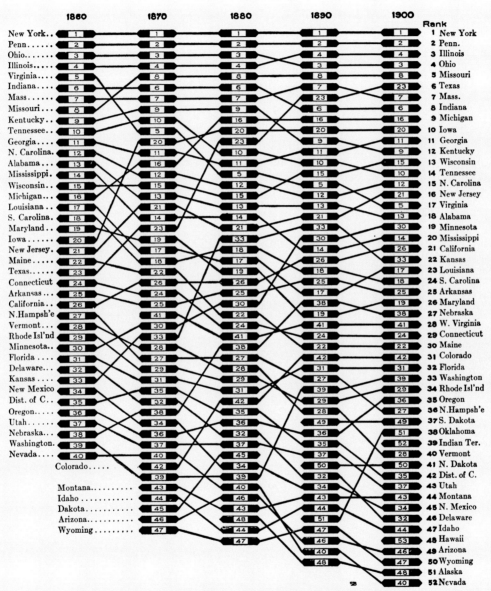

Adapted from the United States Statistical Atlas

ILLUSTRATIVE DIAGRAMS / SYMBOLS I

Symbols, icons, graphic devices – call them what you will – these elements are a vital tool for the information designer. They can be primitive geometric shapes or refined pictograms, or anything in between. They are used to create a visual language, substituting effectively for words, and through their non-verbal form transcend normal language barriers. Sensible colour application gives them even greater scope.

As with all information design, consistency of style is vital if symbols are to function effectively, particularly when several are used together. Clarity of design is also essential – ambiguity or fuzziness of thought in the design process is communicated just as easily as precision. Use the same angles, line weights and colour groups to coordinate the design style for groups of symbols, but be sure that each symbol has a clear identity and is not easily confused with any other.

24

Right Plain geometric shapes serve well as basic symbols. They will reproduce well in print at small sizes. Balance dimensions – don't necessarily make them exact. Many shapes are available to the digital designer as dingbats or typographer's sorts.

Right Combining geometric shapes is the key to simple but effective symbol design. Use the shapes as building blocks to maintain consistent treatment, but don't be too rigid. A little subtlety can add a great deal of visual sophistication. Constraining angles within a group of symbols, as well as maintaining a balance in weight, holds the group together. Remember that positive and negative also work together – space is as vital in devising symbols as it is in designing type.

Symbols have long been widely used as a substitute for written language, particularly when communication is international. Consequently many symbols have specific accepted meanings, and should only be used in that way.

Right Road signs are a good example of geometrically based shapes with precise internationally accepted meanings. Although this limits the broader use of the shapes, it does mean that the icons are known and so need no explanation.

Right Some well-known symbols are subject to copyright law and may only be used with permission. These include the Olympic circles and trademarks such as the Yellow Pages fingers and the London Underground roundel.

Right Many symbols, such as these, have well-established meanings. Although these icons can be modified to suit a graphic style, the general shape must be accepted as a recognized form.

Right Flags are another obvious form of internationally recognized icon. Note that some have complex designs, and flags vary in shape. There are several collections of clip art flags, and at least one source available free on the Web.

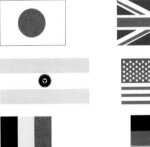

Right and below A functional symbol design must be both legible and unambiguous. Test your design at various sizes, particularly if it is for screen display – remember that users may not have a high resolution monitor. Also, test colour symbols on different backgrounds.

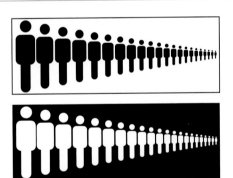

Right Anti-aliasing will help the visibility of small symbols. In these examples the same symbol is shown at a range of sizes with and without anti-aliasing. The enlarged samples show how the process effectively 'blurs' the pixel edges.

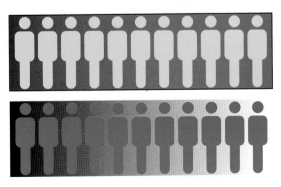
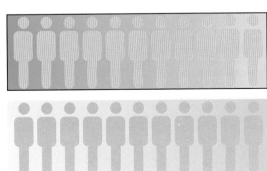

ILLUSTRATIVE DIAGRAMS / SYMBOLS 2

Right The examples of symbol design shown here offer a range of stylization of the human form, from simple geometric shape through to near-illustrative treatment. Notice how successfully each reproduces at various sizes.

Right Here you can see a range of finishes for a symbol design. Digital reproduction and creation methods enable sophisticated use of colour and line, even at small sizes.

Right The basic human-form symbols shown above can be adapted to represent variations of age and gender. Obviously, the treatments need to be consistent in style in order to create a cohesive group.

Right Accessories added to the human form create symbols representing specific activities. Keep the additions simple and in style with the general treatment – a well-placed basic shape will be more effective than a fiddly, complex rendering.

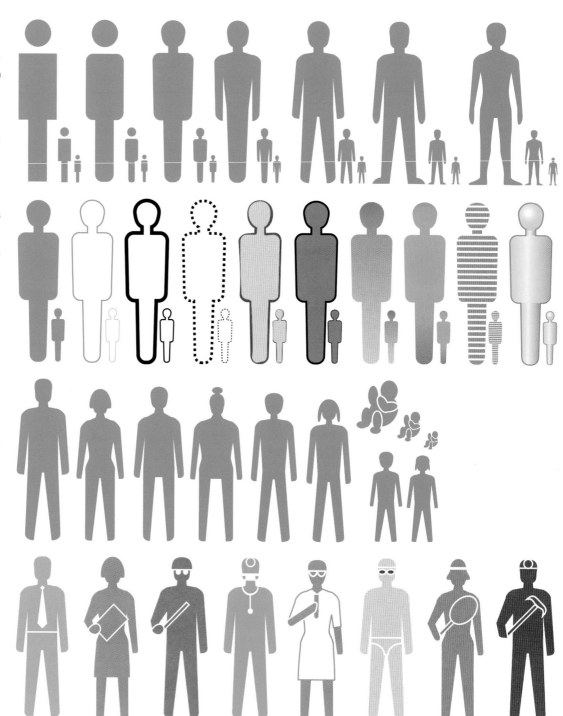

Right Use angles to suggest dynamic movement, as in these sports symbols. Once again, a few simple shapes easily suggest the appropriate accessories, which help identify the activity.

Right Other activities can be indicated by the inclusion of tools and equipment. These activity symbols can represent an entire industrial process. Note the use of the outline shape to hold the symbols as a group.

Right Most human activity involves the hand, and hands are a very good basis for symbol designs. Hand gestures or hands with tools are easily suggestive of precise activity.

Right Symbols work best when designed as a group. Disparate designs, picked out of various clip-art sources, will not hold together within the overall design. Use consistent line weights and outline shapes to provide consistency.

Right Where symbols need to be arranged in groups, use careful alignment with even, balanced spacing. Having a consistent shape for a series of symbols makes aligning and spacing much easier.

ILLUSTRATIVE DIAGRAMS / SYMBOLS 3

Right The process of creating a symbol from photographic reference requires careful formalization and refinement. Make sure the reference is typical and clear. An underlying grid helps the styling process. An alternative visual treatment is also shown.

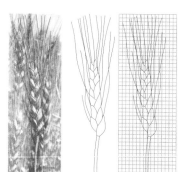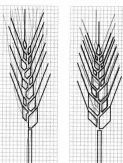

Right Shearing, or skewing, the symbol, either digitally or in construction, allows interesting views other than flat-on profiles. They also give a simple 3-D effect.

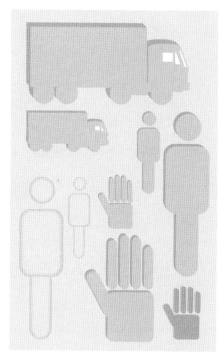

Left Adding an edge to one or more sides produces another 3-D effect. These are particularly effective in 'lifting' the symbol above the page or screen surface. Using software effects such as embossing also creates interesting symbol shapes.

Below Rendering 3-D symbols in tone and colour produces full-blown 3-D objects. These images are virtually complete illustrations, rather than simplified icons, yet still reproduce well at small sizes if carefully designed.

Right Symbols can, of course, be created in 3-D form. These have a greater feeling of solidity, even when in simple, line form. Use a standard grid pattern based on 30°, 45° or 60° angles in order to maintain consistent styling.

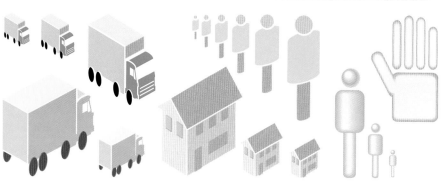

Right Symbols can be incorporated in graphics in a number of ways – maybe as a small icon or as a background image. The symbol can even form the basis for the line graph.

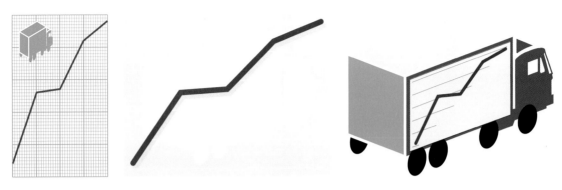

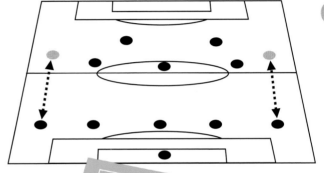

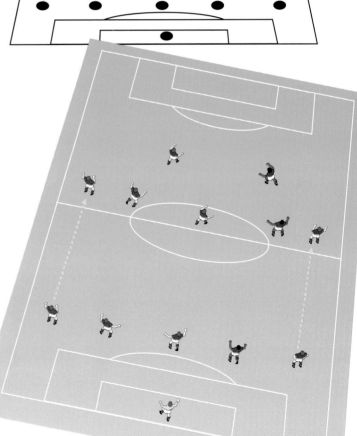

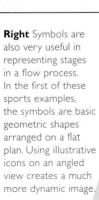

Above Symbols are probably most frequently employed in maps, where they function well as a substitute for labels. Here it is essential that each symbol is clearly distinguishable, and that a key explains the function of the icons used.

Right Symbols are also very useful in representing stages in a flow process. In the first of these sports examples, the symbols are basic geometric shapes arranged on a flat plan. Using illustrative icons on an angled view creates a much more dynamic image.

ILLUSTRATIVE DIAGRAMS / PICTORIAL I

With so many possible variations in style, structure and levels of complexity, in this section it is only feasible to deal with the basic guidelines and approach to illustrative pictorial diagrams. As in all methods of diagram design, clarity is the key factor. This begins with a clear brief and carries through with clarity of thought at every stage. The end user does not have direct access to the reference material, so it is the function of the designer/illustrator to extract the fundamental elements from the brief and reassemble them in a clear and pleasing way, free of any ambiguity, clearly labelled where appropriate and with no obstructions to understanding. With digital technology, specialist 2-D and 3-D drawing programs and the vast selection of available clip art covering most subject areas, stunning and functional pictorial diagrams can be generated even by designers with very limited drawing ability.

Right One step on from symbol designs, basic images such as these simplified human characters and objects serve to illustrate complex activities in a technical brochure. The clean line and simple use of tone also convey the technological nature of the subject matter.

If you need to include labels on illustrations, make sure they clearly relate to the relevant points. Various systems can be used. Leader lines direct the label to the identified point. The leaders can be at random angles, but look neater when conforming to 90° or 45°. Use a broader, white line as a blanking strip behind leader lines that cross the image. Numbers placed on or against items or located by leader lines allow the labels to be arranged as an easily readable list.

focussing knob   diopter correction

prism casing   eye piece

tripod connector

objective lens

diopter correction   eye piece

focussing knob

prism casing

tripod connector   objective lens

1 diopter correction
2 eye piece
3 focussing knob
4 objective lens
5 prism casing
6 tripod connector

(1) diopter correction
(2) eye piece
(3) focussing knob
(4) objective lens
(5) prism casing
(6) tripod connector

Right and below
There are several methods of illustrating elements that would normally be obscured in a straightforward illustration. The cutaway (*1*) leaves off pieces of obscuring parts. Ghosting (*2*) uses transparency to reveal hidden items. The 'exploded' illustration (*3*) separates the various pieces but shows them in relation to their proper position.

Line

Delineated line

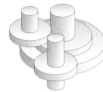

Line and tone

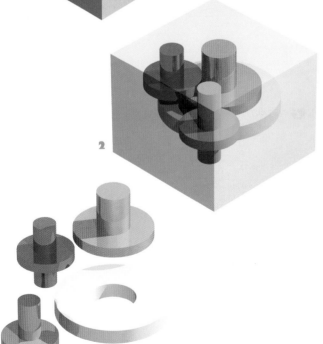

1

2

3

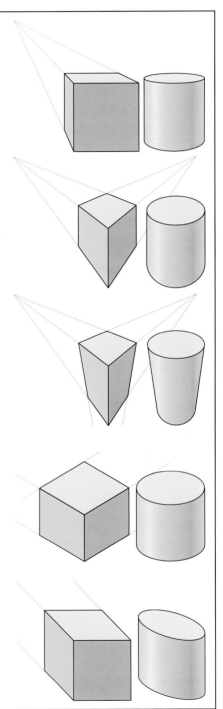

We normally view and draw objects in perspective, with parallel edges appearing to converge as they get further away. Lines projected from these edges appear to meet at what we term a vanishing point, and a perspective drawing may be based on one, two or three vanishing points.

Diagrams can also be drawn in projections that keep sides parallel. This has the advantage of allowing the edges to be measured and so can make drawing easier. To avoid getting too technical, just two types are shown here – isometric and oblique.

A cube drawn in an isometric projection has vertical sides, and the angles of the faces are at 60° and 120°.

Oblique projections have one plane of the cube drawn face on, and the receding planes angled to one side. Any angle apart from 90° can be used, but 45°, 30° and 60° are most common.

ILLUSTRATIVE DIAGRAMS / PICTORIAL 2

Right A variety of graphic styles can be used for illustrative diagrams. These range from simple black-and-white line, through line and tone to detailed realistic treatments.

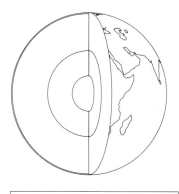
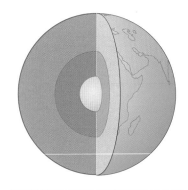
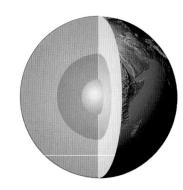

Right Box rules and backgrounds can be useful in giving a regular geometric structure to an otherwise free-form shape. They may be just a basic neutral area of tone or colour, or can form part of the illustration.

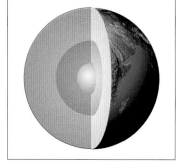
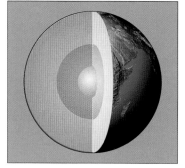
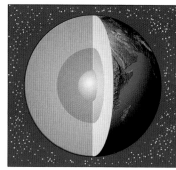

Right There are, of course, many different line styles that can be employed in drawing pictorial diagrams. Choose one that suits the subject matter and audience. While a clean even line suits technical subjects, a varied line is much less formal, perhaps suiting young readers.

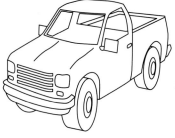

Right There is also a wide variety of colour treatments from which to select. Again, suit the style to the subject/audience. Also, be consistent in style throughout a publication – unless variety is the style.

Right and below
Illustrative diagrams are often used in sequence, breaking down a complex activity into a series of simple steps. Make sure the steps have enough common elements, so that the series flows visually. Label each stage clearly and make sure that the parts for each are separated well enough, so parts don't get confused. In this example, size remains constant up to stage six, when it is enlarged for clarity. Arrows are used to illustrate movement, and their function is explained in a key. Letters are used to orientate the reader, and these are distinguished from the sequence numbers by colour. Note also the subtle use of shading, indicating variations in the planes of surfaces. The sequence is completed by the photograph of the result.

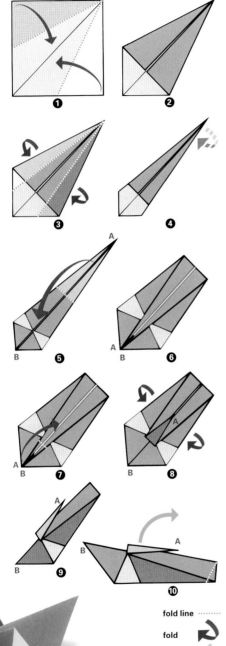

fold line ·······

fold

turn

pull

Right In this diagram, showing how to tie a special knot, colour is used to differentiate each strand of the cord. The arrows not only show where the activity of each step takes place, but also indicate how the cord passes over or under itself or the other strand. The apparent lengths of cord are increased as the diagram sequence builds up. Horizontal rules separate the steps, and each stage is clearly numbered.

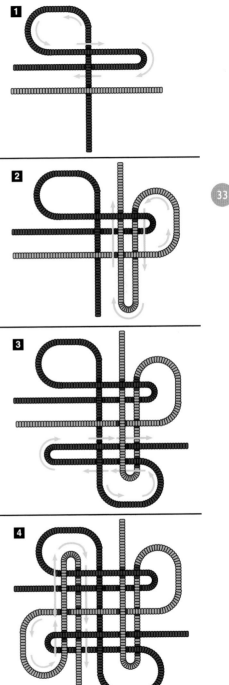

33

ILLUSTRATIVE DIAGRAMS / PICTORIAL 3

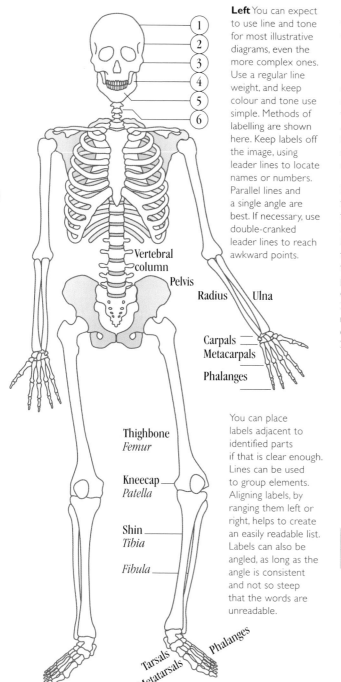

1
2
3
4
5
6

Vertebral
column

Pelvis

Radius   Ulna

Carpals
Metacarpals

Phalanges

Thighbone
Femur

Kneecap
Patella

Shin
Tibia

Fibula

Tarsals
Metatarsals   Phalanges

Left You can expect to use line and tone for most illustrative diagrams, even the more complex ones. Use a regular line weight, and keep colour and tone use simple. Methods of labelling are shown here. Keep labels off the image, using leader lines to locate names or numbers. Parallel lines and a single angle are best. If necessary, use double-cranked leader lines to reach awkward points.

You can place labels adjacent to identified parts if that is clear enough. Lines can be used to group elements. Aligning labels, by ranging them left or right, helps to create an easily readable list. Labels can also be angled, as long as the angle is consistent and not so steep that the words are unreadable.

Right A diagram like this is relatively simple to construct digitally in line and colour. The image is drawn on a 30° projection. It has some basic tone work and graduated tints to give a 3-D effect. Arrows are used to indicate potential movement of parts of the object.

Below Still keeping the detail minimal, and the treatment mostly line and flat tone, this image uses a ghosting technique and arrows to illustrate a simple process. It too is drawn to a regular 30° projection, which makes it simpler to construct using a 2-D vector program.

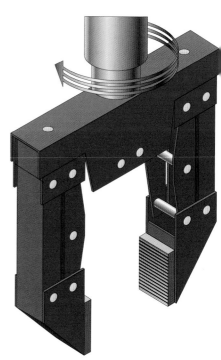

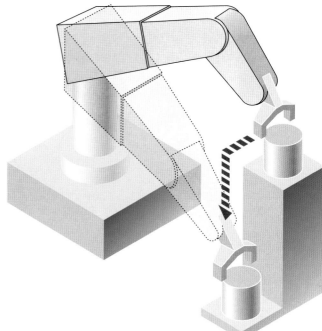

external auditory canal
eardrum
hammer
anvil
stirrup
cochlea
auditory nerve

Round window

Eustachian tube

Left You can give illustrative diagrams for younger readers a less formal treatment. Here the line weight varies to deliberately soften the style. The angled labels also help to liven up what could otherwise be dry subject matter.

Right Particular care is needed when labelling images with lots of contrasting detail. Place a white line behind the black leader, so the eye can follow the line uninterrupted across the illustration. The angular lines also give visual separation from the image.

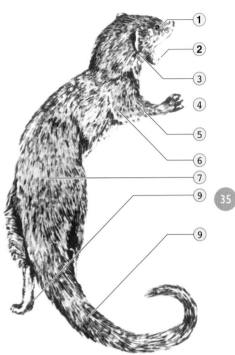

① ② ③ ④ ⑤ ⑥ ⑦ ⑨ ⑨

35

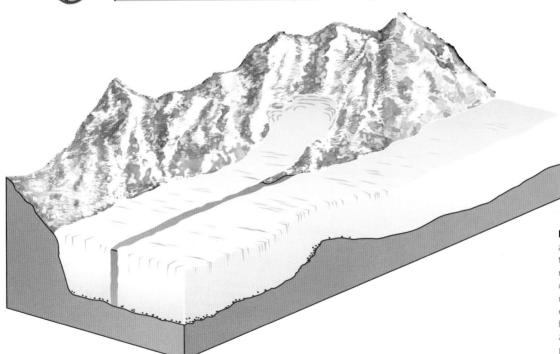

Left A diagram such as this cross-section through a glacier requires competent drawing skill and good reference. It combines elements created in both vector drawing and raster painting programs.

ILLUSTRATIVE DIAGRAMS / PICTORIAL 4

Above Diagrammatic illustrations of reality are best left to the competent specialists. Despite the wealth of ingenious digital aids now available, it takes years of experience to construct images that are convincing. Nevertheless, the current software does facilitate the rough planning of designs to a surprising degree of finish and can save hours of preplanning time.

Left This detailed rendering of a traditional brewing oast house would take some time to draw by any method. Creating it with 3-D software gives the designer the flexibility to select the best view for the overall design. It also enables variations of lighting effect to be tried. The image could be incorporated in an animated sequence. And it could have the further advantage of being available for reuse (subject to copyright conditions).

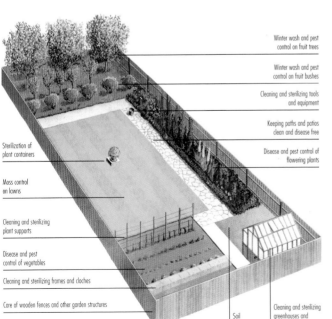

Winter wash and pest control on fruit trees

Winter wash and pest control on fruit bushes

Cleaning and sterilizing tools and equipment

Keeping paths and patios clean and disease free

Disease and pest control of flowering plants

Sterilization of plant containers

Moss control on lawns

Cleaning and sterilizing plant supports

Disease and pest control of vegetables

Cleaning and sterilizing frames and cloches

Care of wooden fences and other garden structures

Soil sterilization

Cleaning and sterilizing greenhouses and conservatories

Left Except for the cut-out shape and the labelling, this diagram is a conventional illustration. This kind of realism would be difficult to create with a 3-D program; and for a sensitive rendering, competent illustrative skills would certainly be required.

Right You may need to create illustrations for cover designs or as more decorative images. An informal assembly of pictorial symbols such as this represents technical subject matter well and is simple to devise.

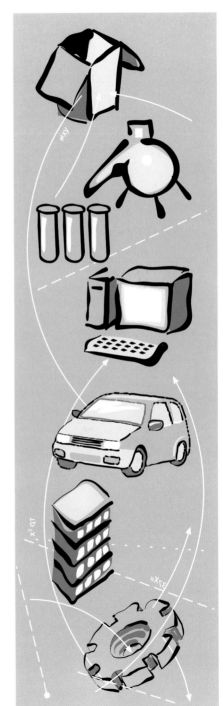

STATISTICAL DIAGRAMS / TABLES I

Presenting data in a clear and unambiguous way is the function of the information designer. Even when data is displayed in tabular form, as a matrix of values, careful planning and preparation is necessary. The eyes of the user need to be guided directly to the salient information, without the interruption of graphic clutter or unplanned confusion.

There are many graphic facilities at the designer's disposal, and intelligent application of these will create tables that are easy to read and therefore clear to understand. As in all information design, consistency of treatment is no less vital than diligent planning.

Tables will, of course, often include text as well as data, while others will contain only text. In order to enhance the appearance and usability of a table, you can incorporate basic graphics, such as small icons or even simple charts.

Right In its simplest form, a table is just a list of information presented in an ordered fashion. This requires the careful alignment of table entries and can be helped by the inclusion of horizontal and/or vertical rules.

Right This basic table sets out the data so that all the vertical columns are equal in size, as are the horizontal rows. The column headings are centred above the relevant entries and are separated from the body of the table by a rule. Value totals are set in bold.

Right Column widths can be varied to fit wide headings or text entries. To carry the eye across the table, 'fill characters' can be used instead of rules; they also prevent the eye being drawn down by a vertical line of repeating entries, such as zeros.

Right Horizontal and/or vertical rules help structure the table. Use rules when tables need to be wide or deep. They help guide the user. Keep the rules as subtle as possible, and leave enough space all round entries.

Right Alignment of type in tables is a matter of style and common sense. Text often looks better centred, while row headings might be better ranged left or right. Decimal points and commas for thousands and millions should be vertically aligned.

thousands	decimal
0,000	0.0
00,000	00.0
000,000	000.0
00,000	00.0
00,000	00.0
000,000	000.0
0,000	0.0
00,000	00.0
000,000	000.0
0,000	0.0
00,000	00.0

Right Instead of rules, bands of tints alternating with white bands can be used behind rows of entries as a means of directing the eye across the horizontal lines of data. Keep the tints subtle, so they don't interfere with the visibility of the entries.

Right Vertical tint bands and rules also help with visibility and assimilation of data. Again, keep tints subtle – otherwise, what should be a background aid will become a foreground nuisance.

Right Alternating bands of colour can also be effective, as long as you don't make them too prominent. If colour is not an option, try two grey values. Or use a constant tint of grey or colour behind the whole table, with white rules as a separator.

Right Embossing and engraving effects can work just as well on columns as on rows. But it may be better to use the fancier treatment for one direction only, combined with a simple rule technique for the other.

Right Effects such as embossing, engraving or even subtle bas relief can work well as horizontal dividers. But bear in mind that what you're trying to do is to facilitate the reading of the table as a whole. Don't let the effects impede.

Right As column widths are likely to be greater than row heights, some subtle effects – such as a drop shadow or a curved or chamfered edge – can be very effective in giving tables a little extra style treatment.

STATISTICAL DIAGRAMS / TABLES 2

Right As part of the function of designing tables, you need to take a look at the way entries are ordered. Often they are in an arbitrary sequence, a result of the data compilation, which is not going to aid the user. Always refer to the client before changing the order.

Belgium	Europe	10.87	54.65	65.52
Estonia	Europe	16.98	68.35	85.33
South Africa	Africa	20.35	34.55	54.90
USA	North America	13.68	29.47	43.15
Mexico	Central America	17.25	45.36	62.61
China	Asia	12.84	75.11	87.95
Australia	Oceania	11.65	36.24	47.89
UK	Europe	9.48	62.87	72.35
Peru	South America	14.32	21.65	35.97
Pakistan	Asia	8.97	63.47	72.44
Kenya	Africa	17.33	47.11	64.44
Germany	Europe	19.56	33.68	53.24

Right More complex table layouts will have subsections or even sub-subsets. Careful typographical layout will help the reader to see these in their correct relationship. Indenting the subsets is effective, but takes up horizontal space.

40

Right Tables supplied in a spreadsheet or similar format can be adjusted easily using available tools. It is worth learning some spreadsheet basics for this. The whole of the table can be reordered so that row headings are in alphabetical order.

Australia	Oceania	11.65	36.24	47.89
Belgium	Europe	10.87	54.65	65.52
China	Asia	12.84	75.11	87.95
Estonia	Europe	16.98	68.35	85.33
Germany	Europe	19.56	33.68	53.24
Kenya	Africa	17.33	47.11	64.44
Mexico	Central America	17.25	45.36	62.61
Pakistan	Asia	8.97	63.47	72.44
Peru	South America	14.32	21.65	35.97
South Africa	Africa	20.35	34.55	54.90
UK	Europe	9.48	62.87	72.35
USA	North America	13.68	29.47	43.15

Right Varying the weights of rules is useful if horizontal space is tight. Make sure the difference in weight is apparent, without making the rules too bold or incongruous.

Right Perhaps one column of data is more important than the others. The table could then be set so that this column ranks entries in ascending or descending order.

South Africa	Africa	20.35	34.55	54.90
Kenya	Africa	17.33	47.11	64.44
China	Asia	12.84	75.11	87.95
Pakistan	Asia	8.97	63.47	72.44
Mexico	Central America	17.25	45.36	62.61
Germany	Europe	19.56	33.68	53.24
Estonia	Europe	16.98	68.35	85.33
Belgium	Europe	10.87	54.65	65.52
UK	Europe	9.48	62.87	72.35
USA	North America	13.68	29.47	43.15
Australia	Oceania	11.65	36.24	47.89
Peru	South America	14.32	21.65	35.97

Right If colour bands are used, subsets can be related by having tints of main colours. Make sure your tint values are enough to be distinguishable, but not so strong that they generate an unpleasant or garish effect.

Right It may be that the final column, the totals or summary, contains the most significant facts. So perhaps this column should determine the ordering of the others.

China	Asia	12.84	75.11	87.95
Estonia	Europe	16.98	68.35	85.33
Pakistan	Asia	8.97	63.47	72.44
UK	Europe	9.48	62.87	72.35
Belgium	Europe	10.87	54.65	65.52
Kenya	Africa	17.33	47.11	64.44
Mexico	Central America	17.25	45.36	62.61
South Africa	Africa	20.35	34.55	54.9
Germany	Europe	19.56	33.68	53.24
Australia	Oceania	11.65	36.24	47.89
USA	North America	13.68	29.47	43.15
Peru	South America	14.32	21.65	35.97

Right Bolder and lighter type variations or the use of capitals are often enough to distinguish subsets, but fancier effects such as row bands with reversed-out type or a drop shadow can be used to highlight headings.

Right Column heads may also have subsets etc. Here, again, careful typographic arrangement will make relationships clear and easily understandable. This arrangement will require adequate space allocation.

Right This method employs tint bands in the vertical columns in order to group subsections. Tints can be kept simple, or you might be able to use subtle treatments to indicate hierarchical levels of subsets.

41

Right If horizontal space is limited, fitting long labels is always a problem. Angled headings are easier to read than vertical type. Allow enough space at the end to fit an angled rule, remembering that its length is determined by the longest entry.

Right Angled heads are also useful when column entries are particularly narrow or in non-data/text tables. The angled column heading fits the narrow measure comfortably and is readable even at quite steep angles.

STATISTICAL DIAGRAMS / TABLES 3

Right Tables that are mostly or entirely text-based will need variable row and column dimensions. Work these to a grid structure, so that depths are based on one, two or three lines etc. The column widths should be determined by the longest entry in each.

Right A table with a variety of entries (for example, text, data and symbols) needs special care in entry alignment. Hanging all row entries from the topmost line is better than centring vertically.

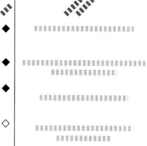

Right Negative values frequently occur in data tables. These can be left to be identified by a minus sign, but if you need to highlight them there are several methods. A simple typographic treatment, familiar to accountants, puts the value in brackets. More noticeable is a variation in weight of type used for negatives, so long as it is visibly different. Use colour for even greater emphasis. Alternatively, the cell containing the value can be picked out by graphic treatment.

	12.3	400	(-23.0)
	(-1.3)	550	25.6
	8.9	231	(-54.5)
	7.3	954	62.3
	3.5	621	(-44.3)
	(-4.5)	333	(-33.7)

	12.3	400	**-23.0**
	-1.3	550	25.6
	8.9	231	**-54.5**
	7.3	954	62.3
	3.5	621	**-44.3**
	-4.5	333	**-33.7**

	12.3	400	-23.0
	-1.3	550	25.6
	8.9	231	-54.5
	7.3	954	62.3
	3.5	621	-44.3
	-4.5	333	-33.7

Right There are times when you may need to pick out other variations in values, particularly in larger tables. Here, too, the use of type weights (so long as the difference is not too subtle), coloured type or graphically differentiated cells will be of service.

Right Add emphasis to selected points in a table by simulating a 'highlighter' effect or a hand-drawn ring or underscore.

	300	300	300
	950	**950**	**950**
	298	298	298
	230	230	230
	254	254	254
	265	265	265
	274	274	274
	269	269	269
	264	264	264

	102	102	102
	274	274	274
	365	365	365
	854	854	854
	1006	1006	1006
	874	874	874
	354	354	354
	655	655	655
	574	574	574

42

Right Where you only need to show a limited range of conditions in a table format, use simple icons with an explanatory key. This will help keep the chart compact and will also give it visual appeal.

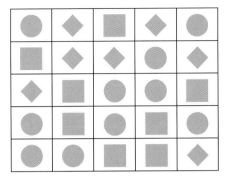

Right Extend the range of possible conditions by varying the size and/or colour of icons. You can also make some of them solid fills and some of them outlines. But don't make so many variations that the user has to struggle to interpret the table.

Right A wide range of graphics can be employed as icons in tables. Using 3-D symbols (such as map pins or shadow) and well-produced pictograms (even flags) will liven up an otherwise staid table.

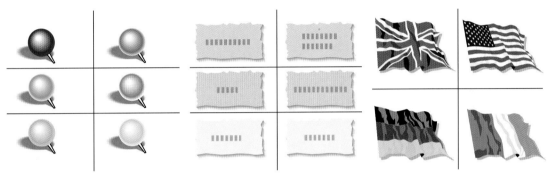

Right Values can be shown in a table format by using multiple or repeating icons. Plan the chart carefully, allowing enough room for the highest value entry and for any text that needs to be included.

Right You can also illustrate values in a table by incorporating a simple bar behind the data in the cell. As the maximum bar length is restricted, subtle differences will not show well.

Right Other types of information graphics can be combined in the table, too. This example shows simple graph and pie charts being used quite effectively, despite their relatively tiny size.

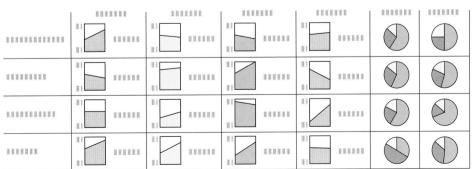

STATISTICAL DIAGRAMS / LINE GRAPHS I

Line graphs are a widely used vehicle for representing data in graphic form. Essentially, they are sequential points of value plotted on a grid matrix with some form of line between the points. The line indicates the rate of change or difference along the progression of the chart: the steeper the line, the greater the rate of change. Line graphs are easy to produce and commonly used, and can be very dull. However, a few simple graphic techniques, applied with intelligence and design sensitivity, can make them very effective communication vehicles.

Bear in mind that the function of the graph is to show the rate of change between points. This is important, because the angle of the line will be varied by the format of the chart – a narrow chart will emphasize the gradient, whereas a wide chart will lessen the effect.

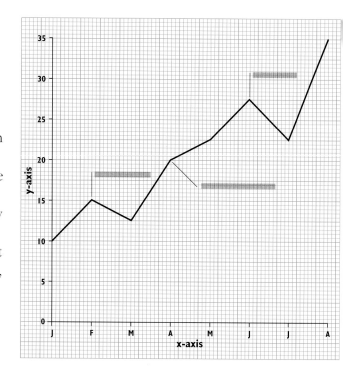

Right A simple line chart with the basic elements identified.

Left Several lines may be incorporated on the same grid. These may be conveniently separate, but if they cross over you must make sure they are quite distinct and can't be confused by the user.

Right Filling in the area between the graph's plot line and the base gives the chart more weight. To emphasize the effect, in this example only the horizontal grid lines have been shown, and these are treated differently above and below the plot line.

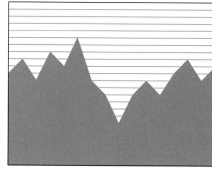

Right Two or more lines may be combined in a single chart but plotted so they accumulate in value. In a graph like this, it must be made clear that the lower line is a subdivision of the total value shown by the top line.

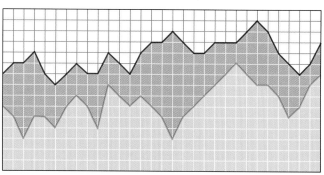

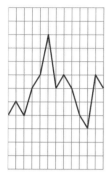

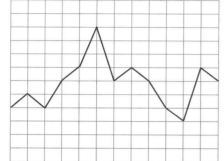

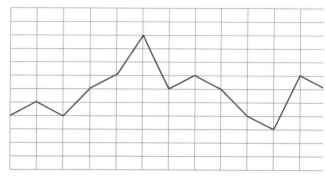

Above These examples show how the angles between points on the plot line are affected by varying the width of the graph – by either compressing or expanding the x-axis while keeping the same depth.

Right Line graphs don't have to be plotted horizontally, although that is the convention. Turning the chart through 90° creates an interesting variation. It also lets you achieve even more dramatic angles on the plot line.

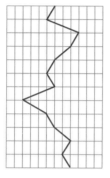

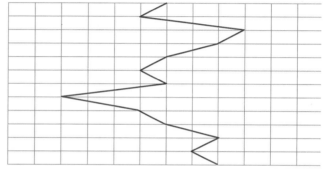

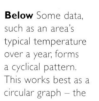

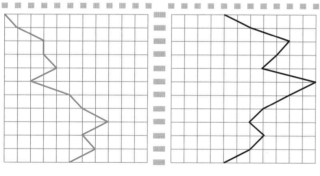

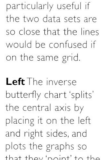

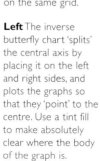

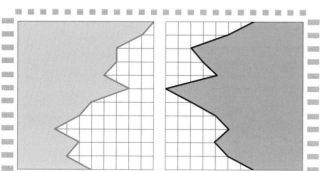

Left As an alternative to plotting two sets of data on one graph, you can plot two charts mirrored so that the base line forms a central axis. This 'butterfly' chart is particularly useful if the two data sets are so close that the lines would be confused if on the same grid.

Left The inverse butterfly chart 'splits' the central axis by placing it on the left and right sides, and plots the graphs so that they 'point' to the centre. Use a tint fill to make absolutely clear where the body of the graph is.

Below Some data, such as an area's typical temperature over a year, forms a cyclical pattern. This works best as a circular graph – the x-axis forming the circle and the y-axis radiating from the centre. As well as making an interesting shape, it shows the repeat process.

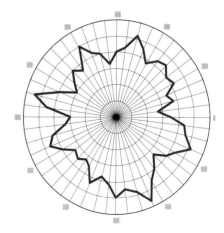

STATISTICAL DIAGRAMS / LINE GRAPHS 2

Right When choosing a suitable line weight for a graph, bear in mind that digitally produced thick lines generate extended mitres, particularly on acute angles, and these may confuse readers. Limit mitre extents with software controls.

Right Point markers can be used to give emphasis to the data points. They can also be used to different-iate lines, as long as there aren't too many lines on the graph.

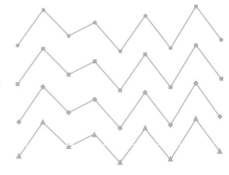

Right Besides line thickness, line style can be varied in order to differentiate multiple lines on monochrome graphs. Make sure the variations in line style are clearly apparent, particularly if the lines are clustered or cross each other frequently. Even when colour is available, you may have to vary the style of the lines so that they are clearly distinguished.

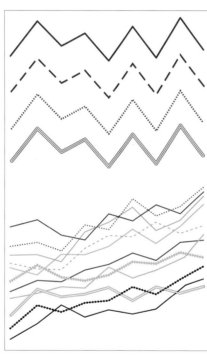

Right Some specialist data sets include high and low or maximum, median and minimum values. On a single line graph, you can have a band to show the range. Or insert markers at the data points.

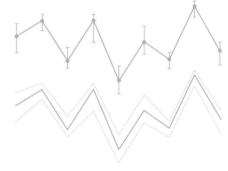

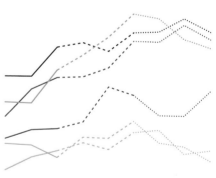

Right This variation on an ordinary line graph has a change in condition at various time points of each data set. This is shown by the change in line treatment. Use a key to explain the reason for the change.

Right Graph styles can be varied by the treatment of the underlying grid. You can either show the grid in full or show just the x-axis and y-axis. Provided that the data points are labelled or easily read, the grid can be left out and the scales shown by tick marks.

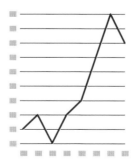
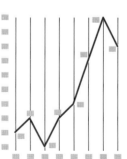
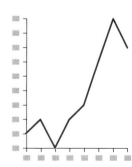

Right You can make
negative values on
graphs more apparent
by colouring the grid
area below the zero
base on the y-axis.
Alternatively, colour
the part of the plot
area that dips below
the line, or change the
colour of the plot line
for the negative part.

Right Sometimes you
may have to combine
two lines that are
plotted to different
axis values. Show
the axes on opposite
sides of the graph,
and use colour to
relate the line to its
axis. It will also help
if the axis labels have
the same colour.

Right Percentage
graphs plot values as
a proportion of one
hundred, with either
side of the plot line
having a value. Make
this clear by using
colour of equal visual
weight for both sides.
This type of graph
often works well if
turned through 90°.

Right When one
value in a data set
varies greatly from all
the others, plotting all
the points on the
same scale will limit
differentiation for all
but the anomalous
value. Various graphic
techniques allow you
to show all the points
without minimizing
the close values.

Right In order to
exaggerate rates of
change, you may want
to use a base for the
graph that is higher
than zero. There are
several ways to make
this clear to the user.
You can leave out
the base line of the
grid, or use one of
these more graphic
techniques.

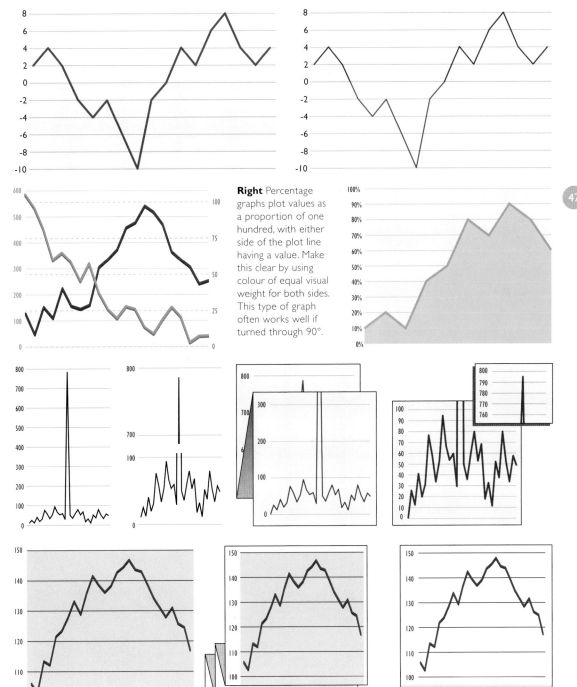

STATISTICAL DIAGRAMS / LINE GRAPHS 3

Right Identifying lines on multiple graphs can be problematic, especially where lines are close together and overlapping. You might have space close enough to each line to label it – but you'll probably need to use various angles, and the result can be messy and unclear.

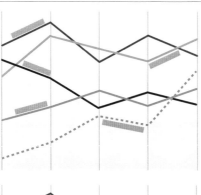

Provided that the line styles are clearly distinguishable (which they should be), use a key listing to identify them. Place the key in a clear area of the graph, so it·doesn't hide any plot lines. Or make the graph narrower and place the key to the side.

Making the graph a bit narrower allows space at the end (or beginning) of lines where they can be named, so long as they are sufficiently separated at the first or last plot point.

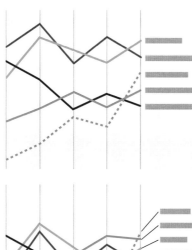

If the lines are too tightly clustered to accommodate adjacent labels, you can use leaders to locate the relevant lines. Otherwise, resort to the key method described above.

Right Whether or not a grid is used, you may want to show the actual values of data points. Position labels so they don't interfere with the plot line. Angling the labels can help. Or use leader lines, so that the values can appear as a row of data.

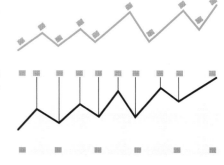

Right If space and style permit, you can have point markers big enough to contain the data labels. An alternative is to use arrow devices that are separate from the line, placed to follow the rise and fall of the plot pattern.

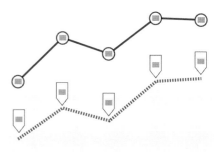

Right and below Key events that have influenced the data on a chart may need to be labelled. The type matter is likely to vary in length and appear at relatively arbitrary points. Use leader lines to locate them. A horizontal plot allows the events to be listed.

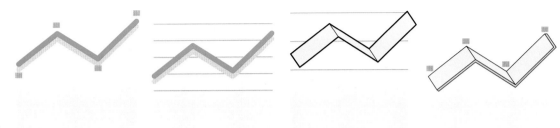

Right Using 3-D effects on line graphs gives a chart a lot of punch. Be careful not to lose or distort the data. The plot line appears to float with a soft shadow, but make sure it relates visually to the grid. These other 3-D plot lines also need careful attention.

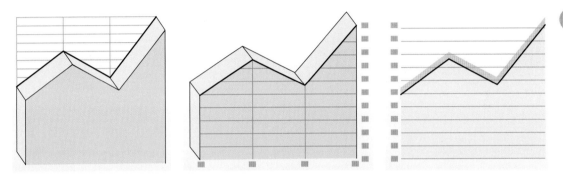

Right You can apply 3-D treatment to filled graphs. This creates a solid block appearance. Here again, make sure any grid lines clearly relate to the plot. The third sample keeps the block very shallow but uses shadow to lift the plot area.

Right Multiple graphs can be placed one behind the other, so long as the plot lines of each graph are visible above the one in front. Make the grid offset alignment clear, too. Here it has been shown only on the face of each plot area.

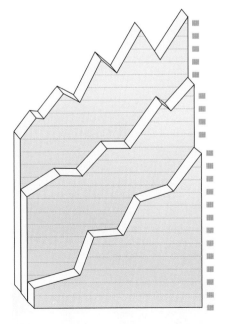

Right Butterfly graphs can also be drawn in 3-D. In order to treat both charts evenly, in this example the view appears to be from above. A viewpoint offset to one side would give emphasis to one or other of the graphs.

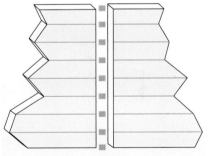

Right This unusual viewpoint and orientation presents the chart as if seen from below or head on. It is very effective in emphasizing the plot line.

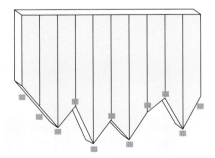

STATISTICAL DIAGRAMS / LINE GRAPHS 4

Right A little creativity can transform even dull charts. This series of charts could have been plotted as one long, ordinary graph. Breaking it up into five-year segments and making a feature of this creates a more interesting effect.

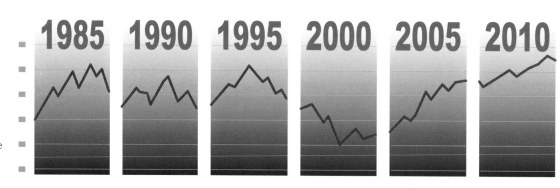

Right These three graphic devices also help to liven up what could otherwise be dull charts, and all are relatively easy digital effects. The first gives the semblance of pins joined by string. An engraved effect is used in the second, and the third employs shooting stars.

Right Incorporating pictorial elements in a chart can be eye-catching. But all too often it is badly done, with the plot line all but obscured. Keep the pictures subtle, and use colour and tone to make the plot area clearly distinguishable from the grid.

Right Imaginative use of chart combinations can create interesting effects. Plotting the lower chart so that it resembles a shadow cast by the 3-D graphs works well when the data for the vertical charts is more closely related than the data for the base chart.

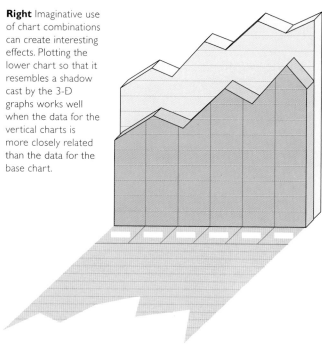

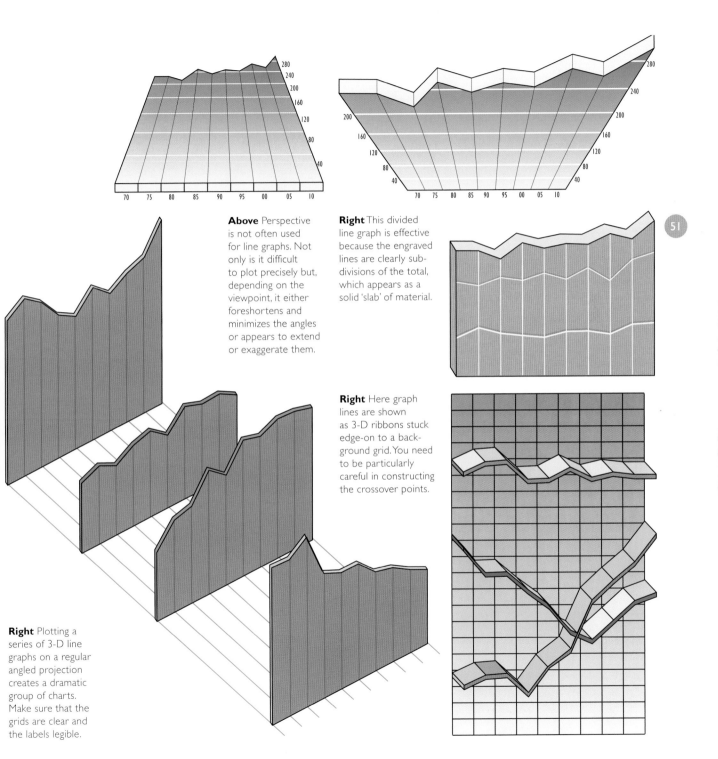

280
240
200
160
120
80
40

70 75 80 85 90 95 00 05 10

200
160
120
80
40

280
240
200
160
120
80
40

70 75 80 85 90 95 00 05 10

51

Above Perspective is not often used for line graphs. Not only is it difficult to plot precisely but, depending on the viewpoint, it either foreshortens and minimizes the angles or appears to extend or exaggerate them.

Right This divided line graph is effective because the engraved lines are clearly sub-divisions of the total, which appears as a solid 'slab' of material.

Right Here graph lines are shown as 3-D ribbons stuck edge-on to a back-ground grid. You need to be particularly careful in constructing the crossover points.

Right Plotting a series of 3-D line graphs on a regular angled projection creates a dramatic group of charts. Make sure that the grids are clear and the labels legible.

STATISTICAL DIAGRAMS / SCATTER GRAPHS

Scatter graphs are a rather specialized type of chart, most often used in scientific and statistical analysis. Like line graphs, scatter graphs are plotted on a grid consisting of an *x*-axis and a *y*-axis – but, whereas line graphs show a progression of points and the rate of change between them, a scatter graph has individual points. The significance is the relationship of each point to all the others on the chart. Configurations of points may illustrate a trend, which can appear as a line.

Right Points on scatter graphs are marked by a suitable device – a small circle or square, for example – which can be in colour. Where two or more sets of data are combined in one chart, each group's points need to be distinguishable.

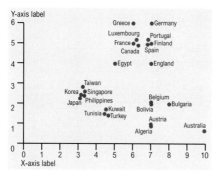

Right When point labels are required, care and patience is needed, particularly when dealing with clusters. Leader lines can be used to locate points. Angled type may be helpful if the configuration permits.

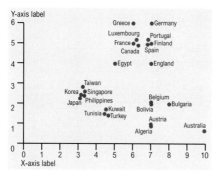

Right It might even be better to use leader lines for all the points, keeping the labels well away from them. Alternatively, a number system should take up less space and can be used across a series of diagrams.

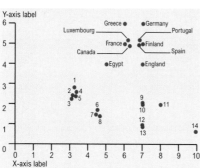

Right This shows the underlying structure of the scatter graph. Points are located by the intersection of two values, one on the *x*-axis and one on the *y*-axis. The axes will usually show two data variables, rather than a chronological sequence along one axis only.

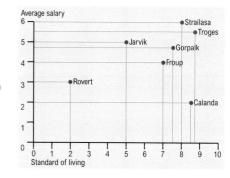

Right A basic scatter graph is simply a configuration of points, often with groups or clusters. If the graph is indicating a general pattern, it may not be necessary to label any of the points.

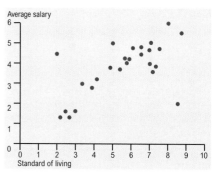

Right This variation of the scatter graph is divided into labelled quadrants. These give a generalized analysis of the conditions pertaining to the points that fall within them. It is best to keep the divisions subtle.

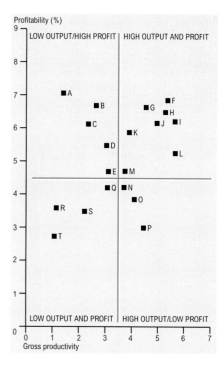

Right Various graphic techniques can be employed to enliven scatter graphs. These include 3-D effects such as embossing and drop shadow, and even more elaborate treatments – such as map pins, pyramids or cylinders.

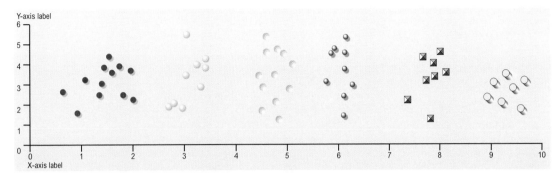

Right Trend lines can be either dominant or subtle, depending on editorial and design preference. In some cases, data for a broad band may be supplied.

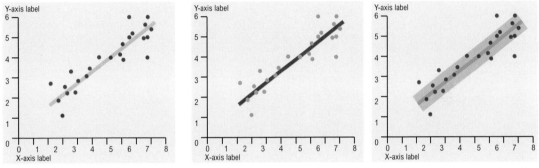

Right A third element of data can be incorporated in a scatter graph by giving the area of the point marker a value. In this case, the colour of the point marker changes according to a keyed scale of values.

Below Another way to display a third value is shown in this 3-D variation of the scatter graph. Here the third value is plotted along a vertical z-axis.

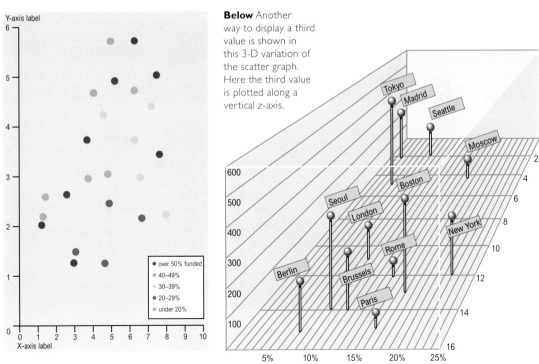

STATISTICAL DIAGRAMS / BAR CHARTS I

Bar charts are probably the most frequently used form of data graphics. Essentially, the bar chart represents values as measurable lengths (the width of the bars remains constant). By placing two or more bars side by side you can see the absolute values of the bars and the relative difference between them. You can draw bar charts vertically – sometimes called column charts – or horizontally. They can of course be drawn at any angle you like, though in order to be read successfully all the bars should extend from a recognizable baseline, so that the bar lengths are easily compared.

The bars are usually labelled individually, although a grid or other scaling method can be used instead. You can plot the bars against a regular scale or a logarithmic one, or they can be drawn in perspective. However, perspective effects do distort and may make the charts hard to read accurately. Bar charts can be enhanced by using 3-D and other effects.

Right In the most basic bar charts, bars of equal width are placed along one axis. Bars can abut or be evenly spaced. The quantity or value is shown by the length of the bar. Typically, each bar shows the amount of something at a different point in time (an annual value, for example); or the bars may show how individual items compare over a given period of time. Labels are usually included to identify bars and for values, although it is possible to use a scale or grid instead of value labels.

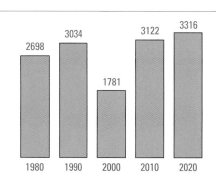

Right Labels can be put wherever you like, so long as they relate to their bar clearly and you treat them all the same way. You can place all the labels inside the bars, or all outside, or do both. In some instances you might want to vary this: for example, you might want to place labels outside short bars, even though other bars have their labels inside. In each case, leave enough space between the label and the edge of the bar. Long labels can be angled, but keep the angle consistent and make sure the label is readable.

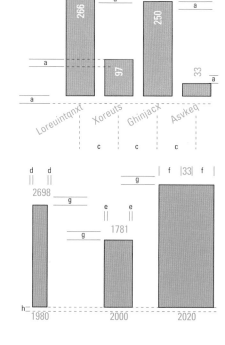

Right Charts can be plotted horizontally, as well as vertically, so that the bar lengths are on the *x*-axis and the incidence of bars along the *y*-axis.

Right A numbered or colour-coded key helps where labels are bigger than bar widths. Place the key in any convenient area of the chart. In a series of charts, be consistent about the position of the key.

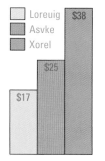
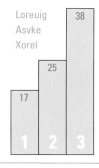

Right The width of
the bars is a matter
of style and common
sense. Narrow bars
may be necessary if
many bars are to be
included in restricted
space. Broad bars give
visual weight.

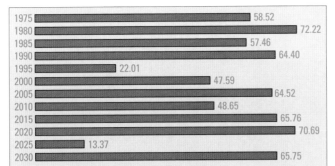

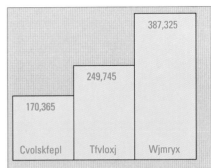

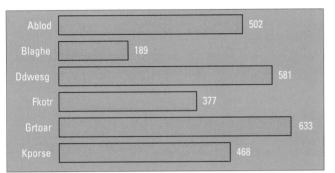

Left and below
If data doesn't have a
set sequence, you can
arrange bars to suit
some other logical or
design requirement.
This could be alpha-
betical order, or they
could be arranged in
either increasing or
decreasing sequence.

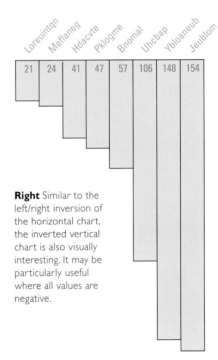

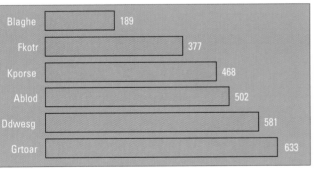

Below Although
horizontal bar charts
are usually aligned
left, transposing the
chart can produce
interesting effects.
Aligning the chart
right also allows you
to range the bars'
identity labels tidily
left, against the ends
of the bars.

Right Similar to the
left/right inversion of
the horizontal chart,
the inverted vertical
chart is also visually
interesting. It may be
particularly useful
where all values are
negative.

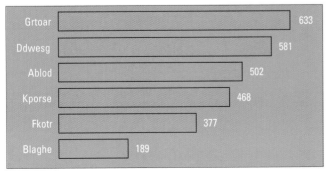

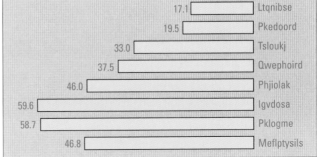

STATISTICAL DIAGRAMS / BAR CHARTS 2

Right Bars don't have to be bars. Any linear structure that has a measurable length can be used – but make sure the style you choose suits the intended audience.

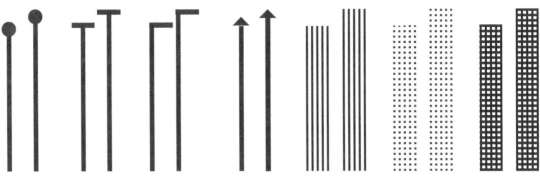

Right You can make bar charts using type characters – either from a regular type font or from dingbats. It's therefore possible to produce charts in a word-processing or page-layout program.

Right Constructing bars from pictograms adds interest. It's best to have the number of symbols relate to the value – not just fitting along the line length. Make each symbol represent one hundred or one thousand items, for example.

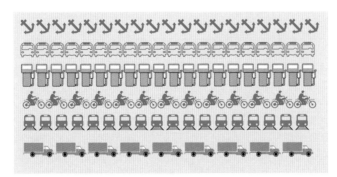

Right Creating fractions of symbols, in order to have them relate closely to the actual value, is simple with a rectangular shape. But with a freer outline (such as the motorbike) you have to exaggerate the smaller fractions, making sure each segment is distinct.

Right Thousands of symbols are available as clip art, but be sure to match styles. Use common sense in the selection (e.g. let derricks relate to oil production, barrels to consumption). Make use of colour to distinguish similar icons that represent different subjects.

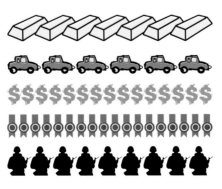

Left Some subjects readily lend themselves to linear symbolism. Transport, for example, with combined aircraft and vapour trail or car and road. But beware of misleading analogy – rail carriages suggest numbers of passengers not travel distance.

Right Charts very often relate to human activity, and lines of human figures are a good way to liven up data presentation. See the symbols section on pages 24–29 for guidance on creating suitable pictograms.

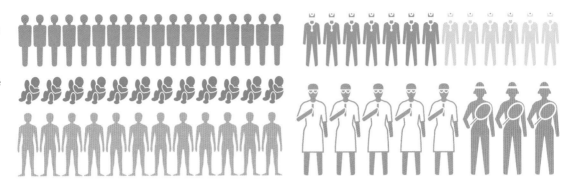

Right This illustrates a common error in charts using human figures. The icons vary in size, and the height represents the value. But the eye sees the overall proportion, which is misleading. For comparison, look at the stacked rows of figures showing the same values.

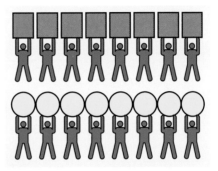

Left If you need to produce a more visible bar form, try combining stronger graphic elements with smaller pictograms.

57

Right Some subject matter inspires an icon that contains the chart, rather than the other way round. Here the content of each test-tube shows the value.

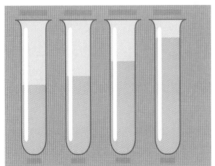

Left A flag on a flagpole presents an identifiable object that can easily be adjusted to show a height value. This is obviously useful for charts comparing states or countries, but the flags could bear company logos or any pictogram.

Right Chocolate consumption and similar subjects lend themselves readily to graphic portrayal. Use your imagination to create eye-catching charts.

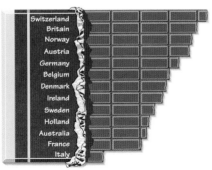

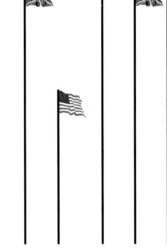

STATISTICAL DIAGRAMS / BAR CHARTS 3

Right You can draw grids for bar charts either instead of or as a supplement to value labels. The grid helps the user measure the bars and differences in bar height. Either draw the grid across the chart's full width (or depth), or just behind bars, or only on the bar front.

Right Instead of a full grid, you may prefer a simple scale alongside the chart, like a ruler, to measure the bars.

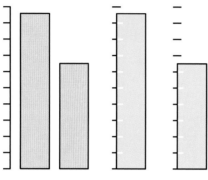

Right Graph paper can make a subtle but effective grid for bar diagrams. Obviously, you should make sure that the graph grid relates to the values of the bars. And it may be too restrictive for a series of charts where you need the flexibility to vary the grid measure.

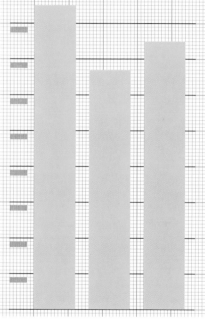

Right Line styles for grids are a matter of taste – or house style. The lines can be solid or broken, and any legible colour. If there are subdivisions on the grid, vary the line weight. Be inventive, while making sure that the grid remains functional.

Right Here are some more variations on grid style. Graduated tones or colour tints can emphasize value differences, with the contrast increasing for the higher bars. A dot structure subtly implies the underlying grid. Also subtle is a pencil effect, with the bar as hard line.

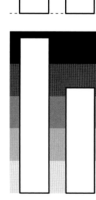
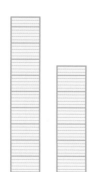
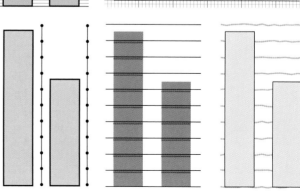

Right Used in more specialist charts, the logarithmic scale helps when there are great differences between values. Logarithmic scales never start from zero. There are several versions of the scale, but base ten is the most frequently used.

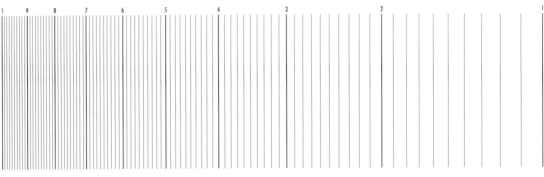

59

Right Perspective scales can start from zero, but plotting bar size accurately is virtually impossible. The structure is based on diagonal lines intersecting at the halfway point. In perspective, the same rule applies.

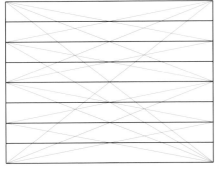

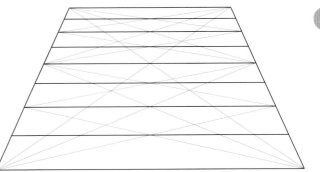

Right and below In some charts it may be necessary to draw the grid to a fixed value – for example, charts based on a 100% total. In other cases a 'benchmark' value may need to be shown. This may vary for each bar in the chart.

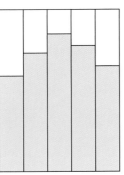

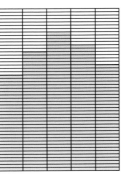

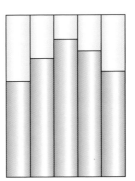

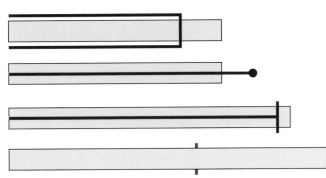

Right Grids can often be incorporated in the structure of the bars. Divide the bar so that each segment equals a scale unit. The final segment may need to be 'cut' to show a fraction. By representing known quantities, pictograms can also be used as grid units.

STATISTICAL DIAGRAMS / BAR CHARTS 4

Right Using 3-D effects can make bar charts much more appealing. But if they are overdone, they can be misleading and/or ugly. Often, a simple offset shadow is enough. This and other basic effects (all of them relatively easy to achieve) are shown here.

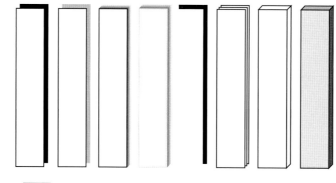

Right This unusual variation of the shadow effect shows the bar appearing to be cut out of the plane of the grid.

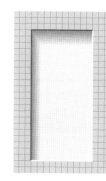

Right Adding a top and side to bars gives them solidity – but don't overdo it. These enhancements are decorative, and it is the face of the bar that carries the real information. Lines all follow the same angle of projection, which should be consistent for a series of charts.

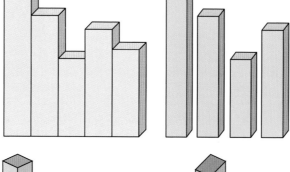

Right Plotting bars with a shallow depth on an oblique or isometric projection makes them appear edge-on to the reader. Nevertheless, the bars accurately represent the values in the data.

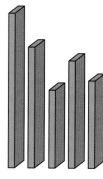

Right Oblique or isometric projections are useful for creating 3-D bars. Angles are constant, and you can either draw the bars on a horizontal base or plot them along the angle of projection to create a more interesting array.

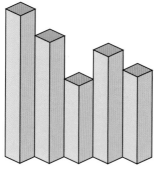
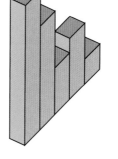
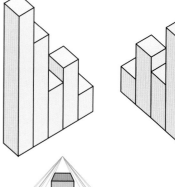

Right Aligning horizontal bars along a central vertical axis creates a pyramid effect. You can keep the angles of the bar ends consistent or draw them to a single vanishing-point.

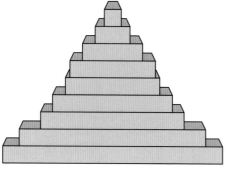
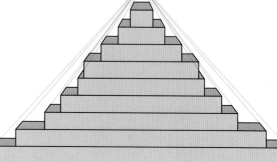

Left You can get another simple 3-D effect by adding an elliptical or circular top and bottom to the bars. The resulting cylinder effect can be enhanced by some subtle shading, or by adding more ellipses at measured intervals along the bar.

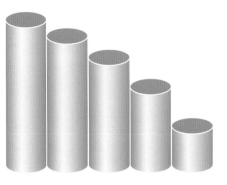

Right You can also plot cylindrical bars along angled axes.

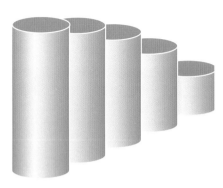

61

Right There are many simple shapes that can be used to create interesting 3-D bars. Just make sure that the visual effect does not interfere with integrity of the chart.

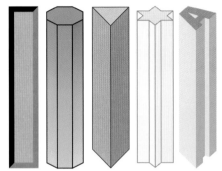

Right Be careful when combining a grid with a 3-D effect. Don't make it hard for the reader to relate the grid to the bar. Wrap the grid around the bar or have it clearly behind. Keeping the bars 'shallow' helps. Or just have the grid on the bar alone.

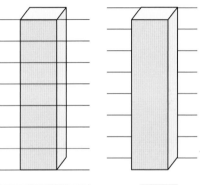

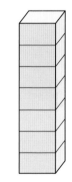

Right Because of its association with the conventional plotting of charts, graph paper makes an effective background. Here it has been combined with another traditional plotting artefact – the map pin.

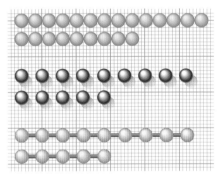

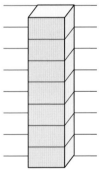

Right Of course, you can create 3-D bars by drawing a series of objects in 3-D. Here, also, shading can be used to enhance the visual effect.

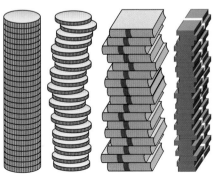

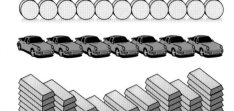

STATISTICAL DIAGRAMS / BAR CHARTS 5

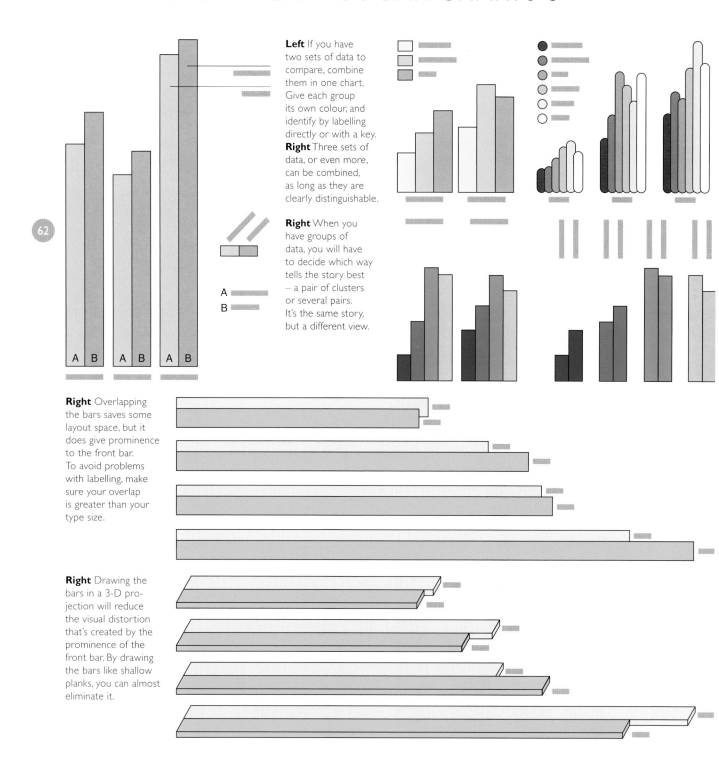

Left If you have two sets of data to compare, combine them in one chart. Give each group its own colour, and identify by labelling directly or with a key.

Right Three sets of data, or even more, can be combined, as long as they are clearly distinguishable.

Right When you have groups of data, you will have to decide which way tells the story best – a pair of clusters or several pairs. It's the same story, but a different view.

Right Overlapping the bars saves some layout space, but it does give prominence to the front bar. To avoid problems with labelling, make sure your overlap is greater than your type size.

Right Drawing the bars in a 3-D projection will reduce the visual distortion that's created by the prominence of the front bar. By drawing the bars like shallow planks, you can almost eliminate it.

62

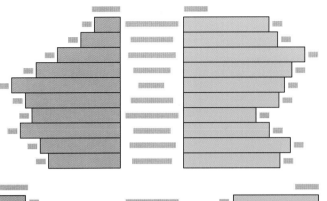

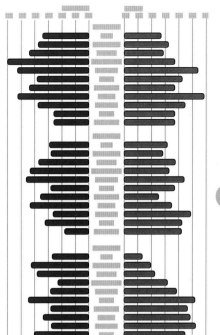

Left Pairs of bars can work well as mirror charts where they are placed either side of a baseline wide enough to include labels.

Right Using a grid to show the values, you can combine several of these charts in a relatively tight space.

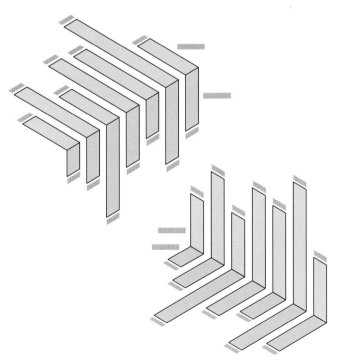

Left This variation on the mirror chart allows more room for labelling – provided the pattern of data suits the arrangement. If need be, you can move the bar indent labels to the left or right side of the chart.

63

Left and right Drawing mirror bar charts at symmetrical angles introduces a dynamic appearance. You'll need to be careful about fitting labels on or around the bars.

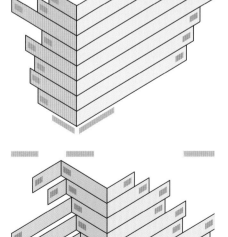

Right This concertina bar diagram combines three charts in an intriguing design solution. The middle series has to run in descending order by value in order to allow the bars of the left-hand chart to be seen clearly.

STATISTICAL DIAGRAMS / BAR CHARTS 6

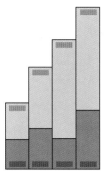

Left and right If you have data for a bar chart showing a total value and breakdown of composite parts, you can produce a stacked or split bar chart. This simply places the component bars end to end, either as a vertical column chart or horizontal bar graph.

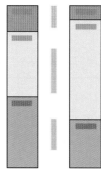

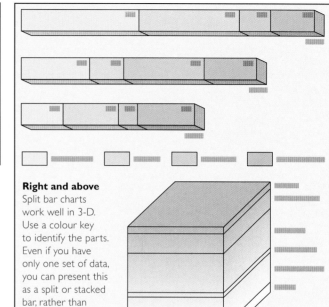

Left Because of the structure, you can plot lines between separated split bars to show rate of change, much as a line graph does.

Right and above
Split bar charts work well in 3-D. Use a colour key to identify the parts. Even if you have only one set of data, you can present this as a split or stacked bar, rather than a regular bar chart.

Left and right
Fitting labels to a split (stacked, composite or divided) bar chart can present problems. Use leader lines to locate labels to parts. Drawing the chart in 3-D and separating the composite parts works well, so long as you make sure the amount of separation is enough to fit in the label. Offsetting the segments also helps to accommodate data labels – as shown here in the stepped split bar chart.

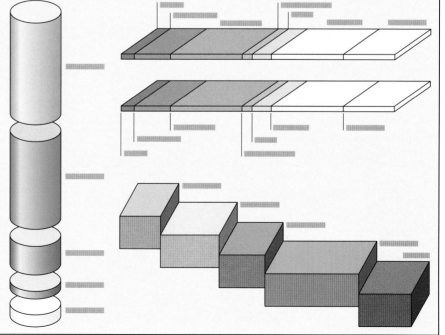

Left and right
Angling 3-D stacked bars creates interest. In particular, the mirrored stacked-bar chart appearing to project from the page surface gives a very dramatic effect.

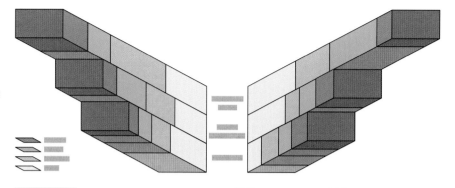

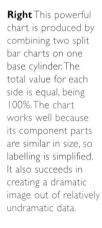

Left and right
If you use symbols or pictograms as the basic bar structure, make sure the shape is as even as possible, to avoid giving undue prominence to any part. The distortion of the base of the test-tube is less misleading than the uneven areas of the human symbol.

Right When the opportunity arises, make use of pictorial enhancement, such as the flags identifying the countries that feature in this chart. The flagpoles become the split bars — and an exciting graphic effect is created in a very simple way.

65

Right This powerful chart is produced by combining two split bar charts on one base cylinder. The total value for each side is equal, being 100%. The chart works well because its component parts are similar in size, so labelling is simplified. It also succeeds in creating a dramatic image out of relatively undramatic data.

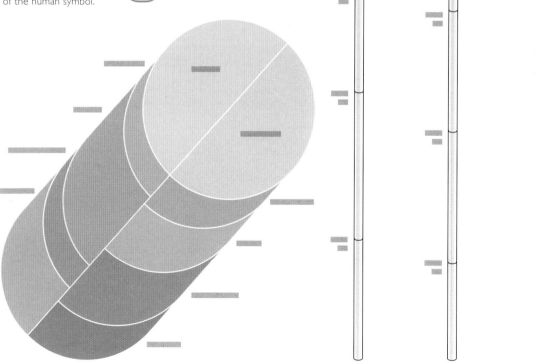

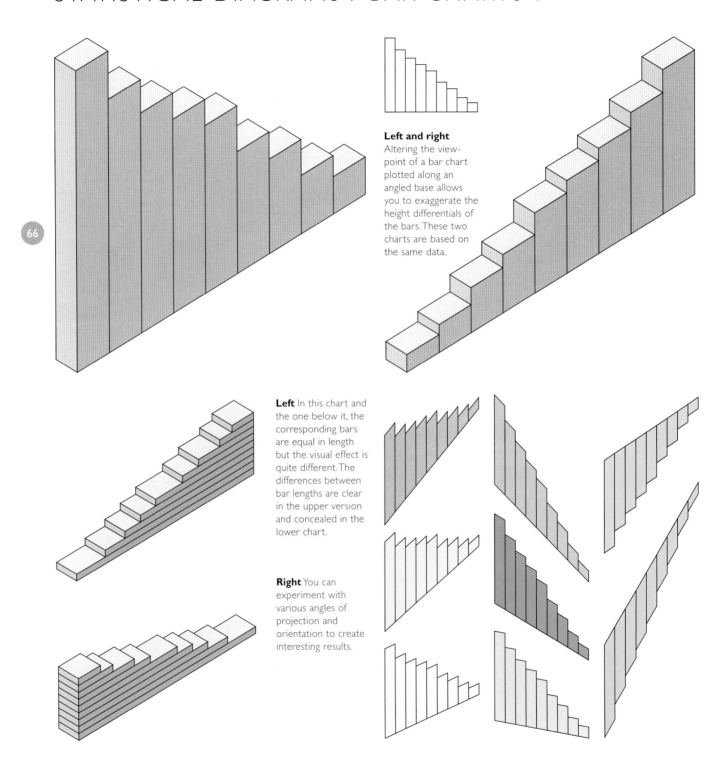

Left and right
Altering the view-point of a bar chart plotted along an angled base allows you to exaggerate the height differentials of the bars. These two charts are based on the same data.

Left In this chart and the one below it, the corresponding bars are equal in length but the visual effect is quite different. The differences between bar lengths are clear in the upper version and concealed in the lower chart.

Right You can experiment with various angles of projection and orientation to create interesting results.

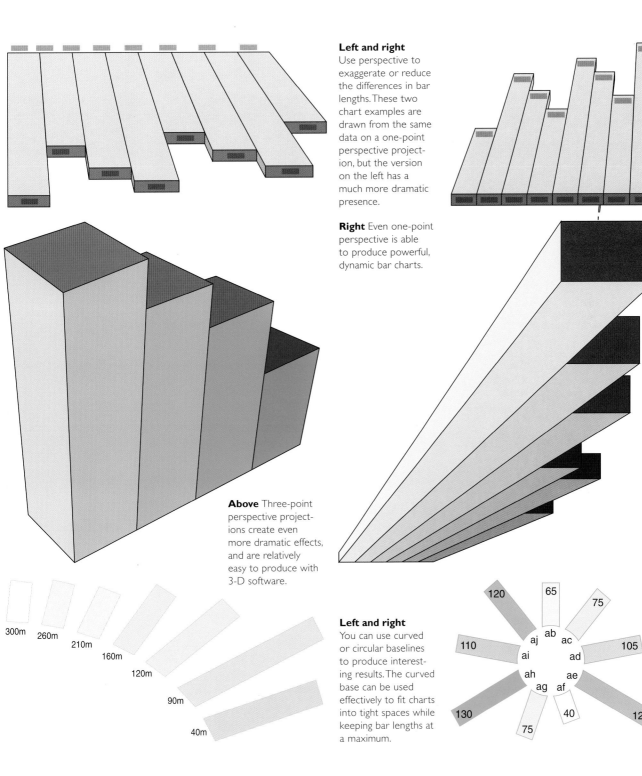

Left and right
Use perspective to exaggerate or reduce the differences in bar lengths. These two chart examples are drawn from the same data on a one-point perspective projection, but the version on the left has a much more dramatic presence.

Right Even one-point perspective is able to produce powerful, dynamic bar charts.

67

Above Three-point perspective projections create even more dramatic effects, and are relatively easy to produce with 3-D software.

300m 260m 210m 160m 120m 90m 40m

Left and right
You can use curved or circular baselines to produce interesting results. The curved base can be used effectively to fit charts into tight spaces while keeping bar lengths at a maximum.

120 65 75
110 aj ab ac 105
     ai   ad
     ah   ae
130 ag af 120
     75 40

BAR CHARTS 8

Right Sometimes charts need to include one or two values that are much greater than all the others. There are several ways of drawing the chart with the excessive bar extending off, while maintaining the visual differentiation of the remaining bars.

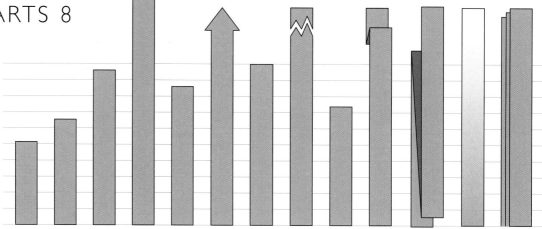

Right If you want to show the complete bar in the chart, you can plot a spiral bar. Construct the long bar to fit the required shape, then draw and duplicate small circles to the bar width. Fit these along the bar length to calculate proportional sizes for the other bars.

Right Similar to the spiral, but much easier to construct, this bar diagram uses curves to highlight the difference in values. Use it when values are not too dissimilar but space for the chart is tight. A grid of equal sized circles aids accurate plotting.

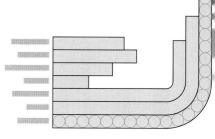

Right This graphic solution effectively employs three charts, one to show the low values and the others for the high values. The apparent folding that connects the charts accentuates the value differences.

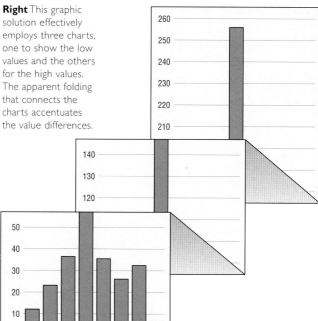

Left An acutely angled projection may be all you need to fit a bar if it is not so much greater than the others.

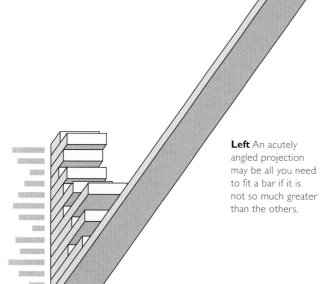

Right There are various methods of displaying negative values clearly in a bar chart. The most common one, shown in this example, has the baseline set part way up the chart, so that the negatives can be displayed as inverse bars.

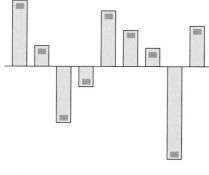

Right If all the values in the data set are high, you may want to have the chart base at a higher value than zero, so that the bar differentiation is enhanced. To make it clear that the chart has a non-zero base, you can draw the bars starting off the chart edge.

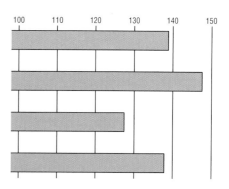

Right In order to maximize the heights of positive bars, you can just leave gaps in the chart where the negative values occur.

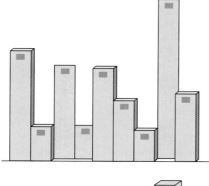

Right A graphic device like this fold effect in the chart also draws attention to a non-zero base.

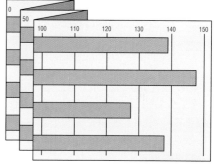

Right This method shows the height of the negative values as flat bars. The advantage is that the negative values are visualized effectively but are still clearly different from the other bars.

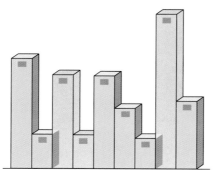

Right This alternative to the one above also shows the heights of the negative values in position within the chart, but here they are drawn to appear as transparent bars.

Right Here we have a folded chart drawn in perspective. This really exaggerates the distortion, while still presenting the data effectively.

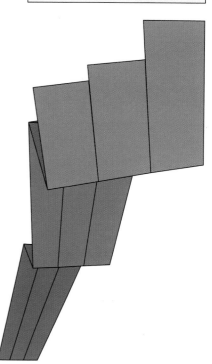

STATISTICAL DIAGRAMS / BAR CHARTS 9

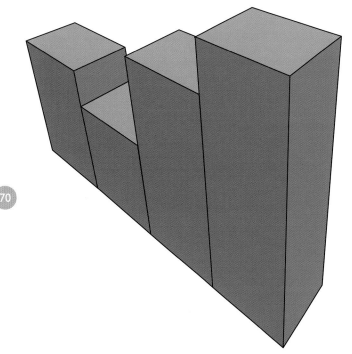

Left Dramatic 3-D effects will give impact to your bar charts. This chart was initially plotted as a basic outline using vector drawing software. Paths were then exported to be manipulated and rendered in a 3-D program.

Right Based on the same data, the outline of this version was drawn in its final shape using vector software. The lines were then imported into a retouching program to be coloured. As it is based on an oblique projection, it has the advantage of being measurable.

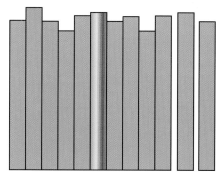

Right Here are several ways in which you can pick out or highlight particular bars in a chart if the general pattern, based on the data, makes all the bars very similar.

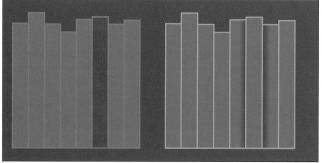

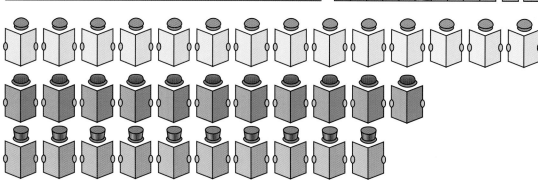

Right You may want to add character to your charts by using pictorial elements that have a more informal, cartoon-like treatment.

Right A chart like this definitely carries visual impact, and is not so difficult to achieve with currently available software, clip art and a little imagination. It could be used to represent a whole range of travel-related activities.

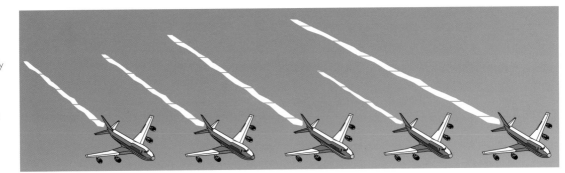

Right Here a familiar object is used as the bar. The cassette is created out of a few circles and lines, and the film strip is even simpler.

Right This example demonstrates how a photographic image placed behind a simple bar chart can create a great effect. Thousands of suitable royalty-free images are now available as clip art.

Right A photograph serves as the base for this chart too. The image was imported into a vector drawing program, and bars made up of the flame icon were imposed in order to create an arresting image.

Major outbreaks

1980
1985
1990
1995
2000

STATISTICAL DIAGRAMS / AREA CHARTS 1

Area charts are most recognizable in the form of the pie chart, so called because it resembles a pie cut into radial slices. Yet there are other forms of area chart, some of them definitely easier to create and label than the familiar pie. Whereas bar charts show values as one-dimensional length, area charts show values as areas. This requires a little calculation. The area of a rectangle is length times width. That of a triangle is half the height times the base. The area of a circle is pi times the radius squared (πr^2). With pie charts we only need concern ourselves with the divisions – converting the total value to 360° and making the angles of the segments divisions of that. Fortunately, spreadsheet and drawing software programs usually have a pie-chart-creating facility. Also, the mathematical equations needed to calculate areas are easy, and with the aid of a computer should present no problem at all.

Left and below Basic pie charts can be designed to contain the segment labels – although you may find labelling a problem when the segments are narrow. Alternatively, draw a smaller pie and place the labels outside.

Left With two or more pie charts to show as a group, you need to decide how to arrange them. If you keep them separate, you should be able to fit labels easily. Try to use the same start point, such as the top vertical, for the first segment in a repeating series.

Below If you overlap pie charts (which has the merit of saving space), make sure you don't obscure too much of the overlaid pies. You can use a key to identify the segments, but leave enough space on the charts to show the value labels.

Right Pie charts are often compressed vertically to create an elliptical outline, which takes up less space. However, this does distort the area and so is potentially misleading, particularly for narrow segments, as is clear from these examples.

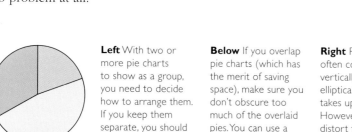

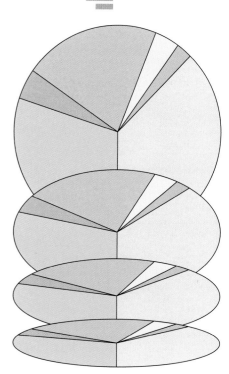

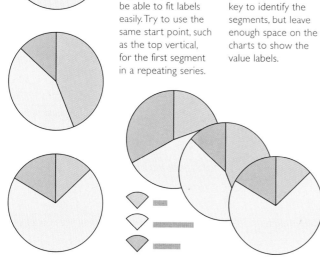

72

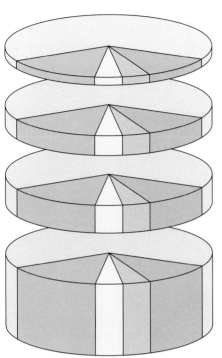

Left Adding a front edge to an elliptical or circular pie chart creates a 3-D effect, but don't make it too deep. The pie is an area chart – and as the edge is added only to the front, segments that are given an edge will have a greater visual area, especially when the same colour is used. Making the edge colours markedly darker or using a different colour will lessen the distortion.

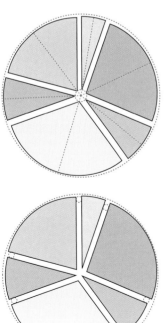

Left Separating the segments to create an 'exploded' pie is a popular technique. As the pie segments will usually vary in size, moving them an equal amount radially out from the centre will produce uneven gaps between segments. However, keeping an even separation means that segments have to be moved varying distances from the centre. The choice is yours.

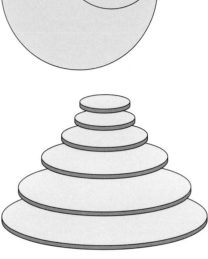

Left Variable circular area charts can also be divided. In this series of overlapping pie charts, the size of each pie represents its total value relative

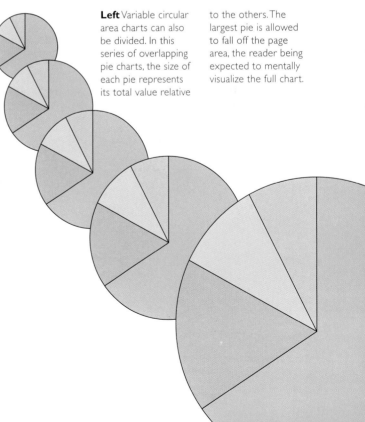

Left Although circular area charts usually take the form of a divided pie, you can use circles with a variety of areas to create another form of area chart. The area of a circle is πr^2. Divide a value, say 42, by π (3.142) = 13.37. This is r^2, so find the square root of 13.37 (which is 3.65) and that's the radius of a circle of area 42. Work out the radii for all the other items in your data set and draw all the circles. You can enlarge or reduce the circles as a group, maintaining their proportions. You can also resize them in one dimension to create ellipses.

to the others. The largest pie is allowed to fall off the page area, the reader being expected to mentally visualize the full chart.

73

STATISTICAL DIAGRAMS / AREA CHARTS 2

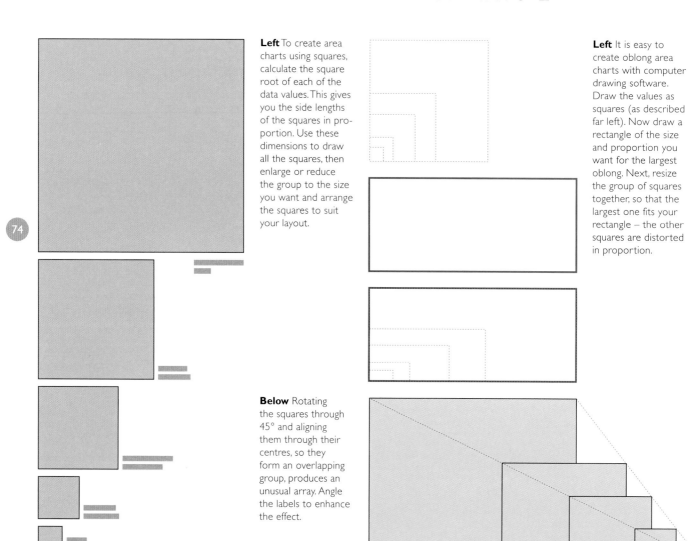

Left To create area charts using squares, calculate the square root of each of the data values. This gives you the side lengths of the squares in pro-portion. Use these dimensions to draw all the squares, then enlarge or reduce the group to the size you want and arrange the squares to suit your layout.

Left It is easy to create oblong area charts with computer drawing software. Draw the values as squares (as described far left). Now draw a rectangle of the size and proportion you want for the largest oblong. Next, resize the group of squares together, so that the largest one fits your rectangle – the other squares are distorted in proportion.

Below Rotating the squares through 45° and aligning them through their centres, so they form an overlapping group, produces an unusual array. Angle the labels to enhance the effect.

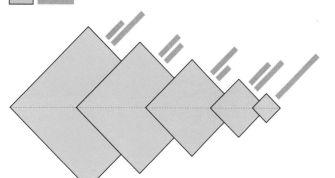

Above and left Overlapping the rectangles so they fit available space is fine, as long as you leave adequate space for labels. Using a hard or soft shadow helps to separate the shapes visually.

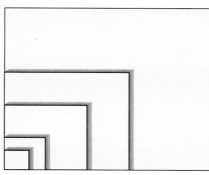

Right Having created your rectangles to the correct proportion, you can use digital techniques to adjust the shapes – by shearing, for example. Experiment with unusual arrangements to make the charts stimulating.

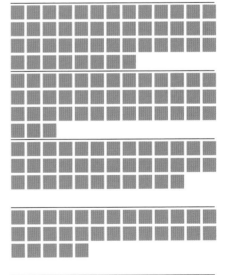

Right This effective (but false) perspective was created in a vector program. After being rotated, each rectangle was resized on the vertical axis only and by varying amounts. The visible areas are distorted, but for conveying a general idea this is acceptable.

Below Even regular flat squares can be set out interestingly. The arrangement shown here works well when the chart has to be integrated with a text layout.

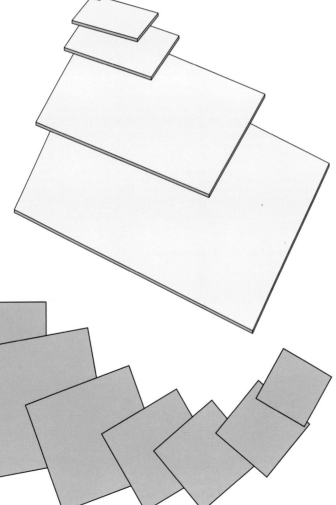

Right Even triangles can be used as area charts. Simply create square areas, using the square roots of the data values. Now delete one corner point of each square to make the triangles and resize them as a group, in proportion, to produce the shape you want.

Below Instead of using a single shape to represent a value, you can use multiple units. This type of diagram, a unit area chart, gives you flexibility in arranging the pattern for each entry.

75

STATISTICAL DIAGRAMS / AREA CHARTS 3

Left In this chart the various areas have been shown against a common benchmark value that remains a constant size. The subject matter is sea area as a proportion of the land areas of islands, but there are other applications.

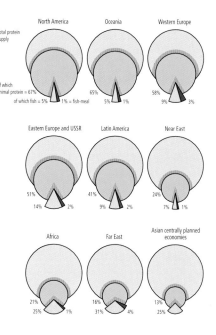

Right This diagram combines circular area charts with divided area (pie) charts. The rearmost circle represents the 100% total for each region.

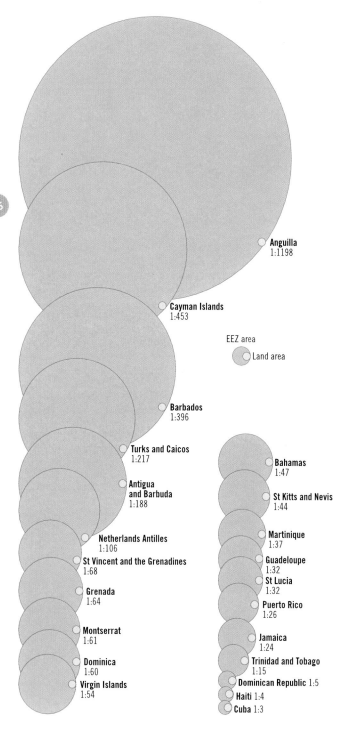

EEZ area
Land area

Anguilla
1:1198

Cayman Islands
1:453

Barbados
1:396

Turks and Caicos
1:217

Antigua
and Barbuda
1:188

Netherlands Antilles
1:106

St Vincent and the Grenadines
1:68

Grenada
1:64

Montserrat
1:61

Dominica
1:60

Virgin Islands
1:54

Bahamas
1:47

St Kitts and Nevis
1:44

Martinique
1:37

Guadeloupe
1:32

St Lucia
1:32

Puerto Rico
1:26

Jamaica
1:24

Trinidad and Tobago
1:15

Dominican Republic 1:5

Haiti 1:4

Cuba 1:3

Right Juxtaposing two half pies that vary in size according to their totals is a neat way of stressing the comparison and saving space, too.

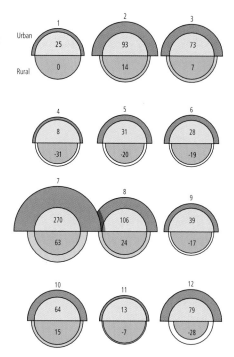

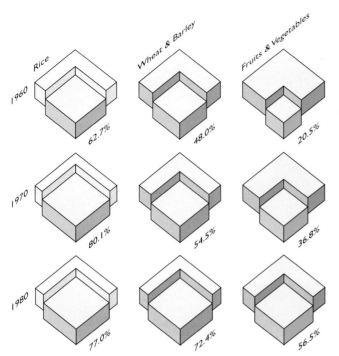

Rice
1960
62.7%
1970
80.1%
1980
77.0%

Wheat & Barley
48.0%
54.5%
72.4%

Fruits & Vegetables
20.5%
36.8%
56.5%

Above In each of the charts in this series, percentages are shown as sections of the full area. You can add interest to a chart drawn from a bland set of data by using an unusual projection and then offsetting a segment, in this case vertically.

Right Royalty-free photographic images can be dropped into rectangular, circular or other shapes used for area charts. But make sure they are still clearly charts and aren't mistaken for odd-shaped pictures.

23% 27% 18% 32%

Right Rectangular area charts can include subdivisions. These split area charts are easier to construct than pie charts, and can be just as effective. Don't overlap charts like these, as the labels could easily be misread.

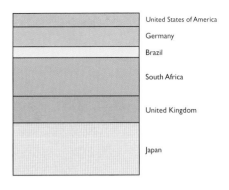

United States of America
Germany
Brazil
South Africa
United Kingdom
Japan

United States of America
Germany
Brazil
South Africa
United Kingdom
Japan

Right Angle split area charts to create a more interesting layout. Construct the charts initially as basic rectangles, then use a shearing process in the vector drawing program to achieve the angle you want.

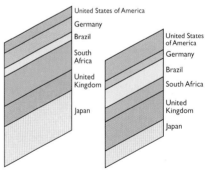

United States of America
Germany
Brazil
South Africa
United Kingdom
Japan

United States of America
Germany
Brazil
South Africa
United Kingdom
Japan

Right This variation of the split area chart works when there are three conditions to be shown, one of which can be treated separately. In this case the 'not sure' value is isolated so that the 'yes/no' comparison is more easily seen.

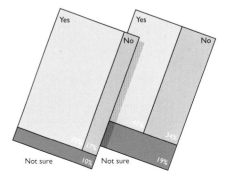

Yes
No
Not sure

Yes
No
Not sure

STATISTICAL DIAGRAMS / VOLUME CHARTS 1

Volume charts are slightly more complicated than area charts to calculate, but they are particularly useful when comparing data vastly different in value.

A bar might be drawn to a width of say 10cm; an area 10cm square contains 100cm²; but a cube measuring 10cm × 10cm × 10cm holds 1000cm³. So long as it will be recognized as a cube, it can represent ten times the value conveyed by the area chart without taking up much more page space.

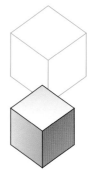

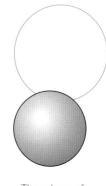

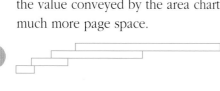

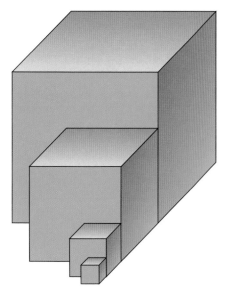

Left These examples demonstrate how volume charts are able to represent vastly greater values in a compact space. Bar charts use only one dimension to express value, and these bars equate to (from the top down) 40, 25, 10 and 5 units. Square area charts represent values as a multiple of width × height, and in this case show 1600, 625, 100 and 25 units. Because the volume chart uses three dimensions (width × height × depth), these four cubes represent values of 64,000, 15,625, 1000 and 125 respectively.

Right Choose the projection for volume that suits your design requirements best. These examples are based on angles of 30°, 45° and 60°. You can also use any oblique projection.

Above These three shapes – the cube, the sphere and the cylinder – all have calculable volumes, which makes them suitable for volume charts. The cube and the cylinder can be drawn convincingly in line, but all three benefit from some careful shading.

The volume of a cube is: width × height × depth.
That of a sphere is: $(4 \times \pi \times r^3) \div 3$.
And the volume of a cylinder is: $\pi \times r^2 \times$ height (or length).

For most general calculation, $\pi = 3.142$.

The cylinder and the cube are the easiest volumes to calculate and draw, but the sphere is also used. Its most common application is in showing the comparative sizes of planets and stars.

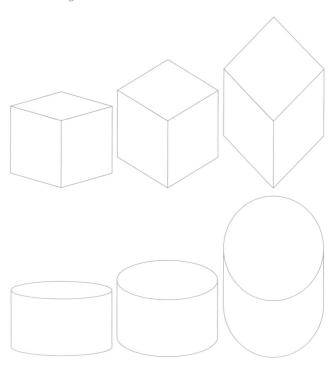

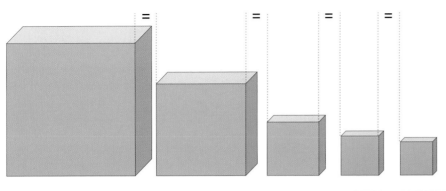

Left You will need to give some thought to the best arrangement for a series of volume blocks. Bear in mind where labels will fit. You can abut the blocks or space them evenly, aligning them at either the front edge or back edge.

Below Volume charts can create striking images – particularly when given imaginative colour treatment, as in this example. The underlit effect adds drama to a powerful structure.

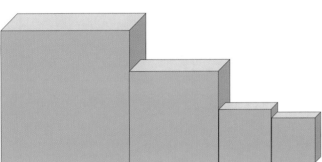

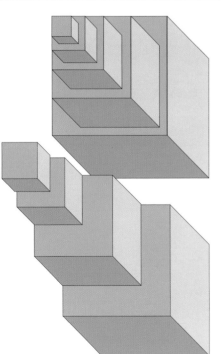

Left Volume charts can be set out in a wide variety of ways, so experiment with various methods. If you choose to 'stack' the blocks, you could keep the chart compact by aligning the top and back edges so that they fit within the top face of the base block. Otherwise you can place them so the base (obscured) corners align. But remember to leave sufficient space for labels.

STATISTICAL DIAGRAMS / VOLUME CHARTS 2

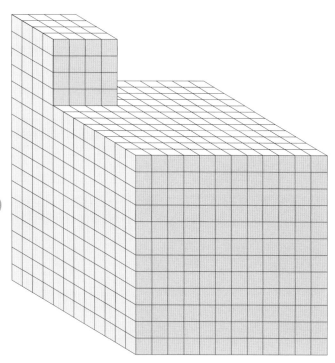

Left Volume charts can be made up from even-sized cubic units. The unit-volume chart thus produced has the advantage of a grid enabling the user to measure values. The charts are simple to construct in a vector drawing program. Just draw one small cube, then duplicate it to build the larger cubes. You'll find that using a regular angle, such as 30° or 45° degrees, is helpful.

Right A grid base makes a very effective background to these cube structures.

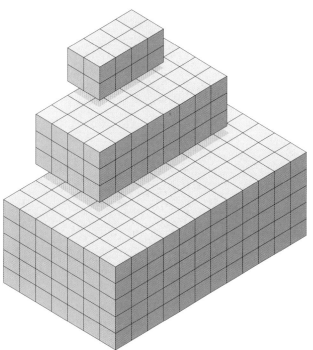

Left You don't have to stick to an overall cube shape for the unit-volume chart. In fact, provided you are working with small numbers of cubes, the style is very flexible and allows you to build shapes to suit your needs.

Right Unit-volume blocks drawn to an isometric format are here laid out in a random pattern conforming to an isometric grid base, to add interest to the design.

Right Since the blocks have a built-in measurable grid structure, you don't have to use complete blocks. You can take out one or more of the individual units. This not only makes an interesting shape, but can be a very useful way of drawing attention to the differences between data values. The upper example is based on a complete cube with blocks removed. In the lower chart, the basis is a linear group. The removal of blocks in the subsequent groups is done in a systematic way, so as to facilitate understanding.

81

Right Using 3-D software, you can produce eye-catching arrangements of unit-volume charts. Just make sure that the reader can see enough of the blocks to make an adequate visual comparison.

Right Other shapes that you can employ to make unit-volume charts include the cylinder and the sphere, which are then built up to form essentially cubic shapes – since making a cylinder or sphere of the units would be asking too much of the reader.

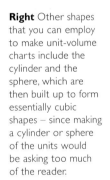

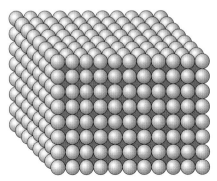

STATISTICAL DIAGRAMS / VOLUME CHARTS 3

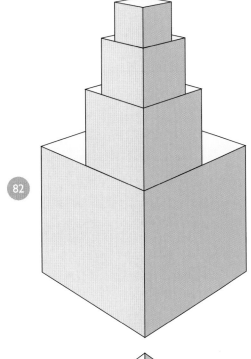

Left Volume charts plotted using a two-point perspective projection – in which the vertical sides remain parallel – have an impressive solidity. Use a 3-D drawing program to create them, as they would be difficult to create accurately using standard 2-D vector drawing software. The sides of the cubes can be used to display an identifying graphic element – such as a pictogram of the subject, or a map.

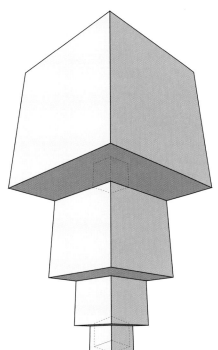

Left Three-point perspective produces even more convincing shapes, but you have to be careful not to make the view too extreme. Otherwise, users will not be able to make a sensible comparison of data – which is the point of the chart. The degree of distortion is indicated by the red dotted outline. This shows how the lowest cube would appear if plotted on the plane of the top cube. Charts like these need to be created in a 3-D program, to obtain correct proportions.

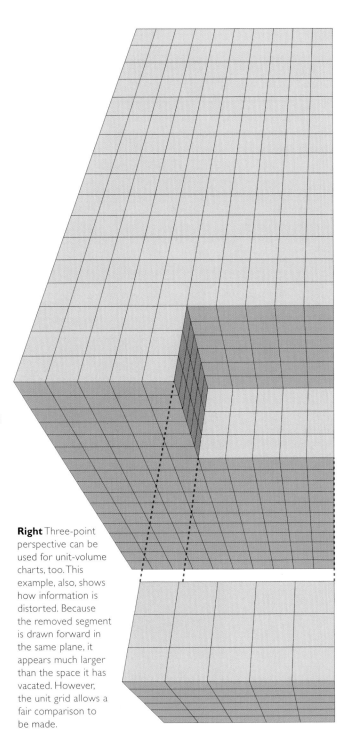

Right Three-point perspective can be used for unit-volume charts, too. This example, also, shows how information is distorted. Because the removed segment is drawn forward in the same plane, it appears much larger than the space it has vacated. However, the unit grid allows a fair comparison to be made.

Right A volume chart with the blocks in perspective and randomly oriented creates interest – but if the blocks overlap, you can create the wrong impression of size. The cubes in each of these three examples are the same size as their counterparts in the other two charts. Notice how the overlaps make the cubes appear to vary in size. In the bottom example the grid provides the reader with a visual control mechanism, making comparison and assimilation of value easier.

Below In this chart showing sizes of stars, the largest value is much greater than all the others. Had the chart been designed to show the largest volume in its entirety, the others would have been minute and indistinguishable in size. Because most of the largest volume has been allowed to fall off the edge of the page, the others appear at a useful size. The curve of the edge of the biggest volume is sufficient to suggest its massive scale to the reader.

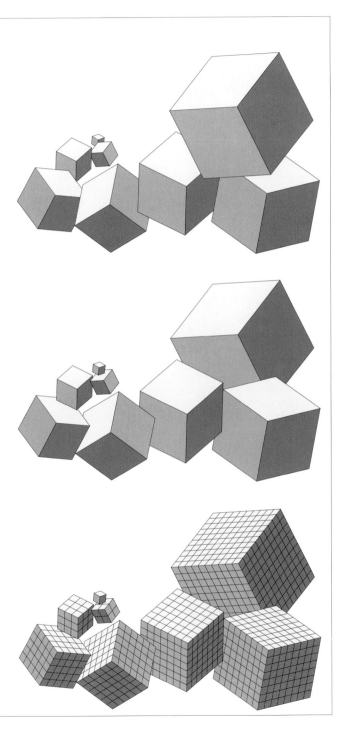

83

STATISTICAL DIAGRAMS / COMBINED CHARTS I

In order to introduce some dynamism into a relatively dull array of charts, you can combine two or more different chart types. Almost any statistical diagrams can be successfully combined, with a little imagination and ingenuity. Make sure that any colour-coded identity is maintained across the chart styles and that scale grids cannot be confused.

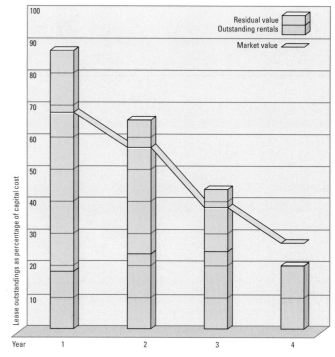

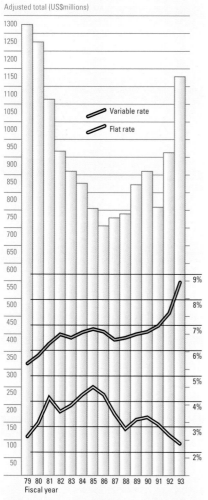

Adjusted total (US$millions)

Right Combining one or more line graphs with a bar chart is quite straightforward, especially if the scale grid applies to all the elements in the chart. Make sure you clearly label each line or bar – a key is particularly useful for this.

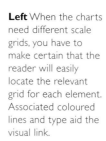

Variable rate

Flat rate

Left When the charts need different scale grids, you have to make certain that the reader will easily locate the relevant grid for each element. Associated coloured lines and type aid the visual link.

Right In this design, unit-area charts are combined with the labels of a simple split bar chart. The leader lines serve both to link the labels to the bar charts – very useful for the shallow bar segments – and to provide a matrix for the unit-area charts.

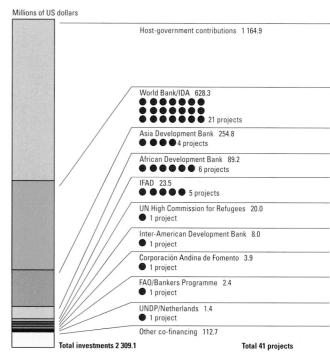

Millions of US dollars

Host-government contributions 1 164.9

World Bank/IDA 628.3
21 projects

Asia Development Bank 254.8
4 projects

African Development Bank 89.2
6 projects

IFAD 23.5
5 projects

UN High Commission for Refugees 20.0
1 project

Inter-American Development Bank 8.0
1 project

Corporación Andina de Fomento 3.9
1 project

FAO/Bankers Programme 2.4
1 project

UNDP/Netherlands 1.4
1 project

Other co-financing 112.7

Total investments 2 309.1   **Total 41 projects**

Right You can use secondary charts to show subsets of data. Here the vertical 3-D bar chart is the dominant element, and the 2-D horizontal charts – which supply the breakdown of information – are colour-coordinated for easy reference.

Below The angled arrangement of these bar and area charts helps hold each group together as a unit. This allows you to use the colours to distinguish the type of chart content, rather than for identifying each country.

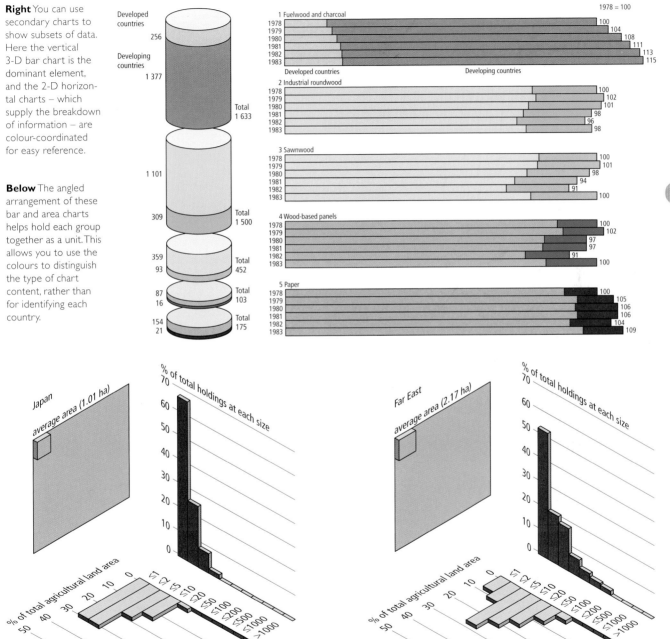

85

STATISTICAL DIAGRAMS / COMBINED CHARTS 2

Right You can show the makeup of one segment of a pie area chart like this, by extending the depth of the segment. In fact, two split column charts are combined with each segment to show the breakdown of the segment value by two methods.

Far right This chart combines a pie and bar diagram, too, but here each segment is given a column value. Note that the data forming the height values is different from the segment proportions. Arrange the chart so that the tallest columns appear at the back.

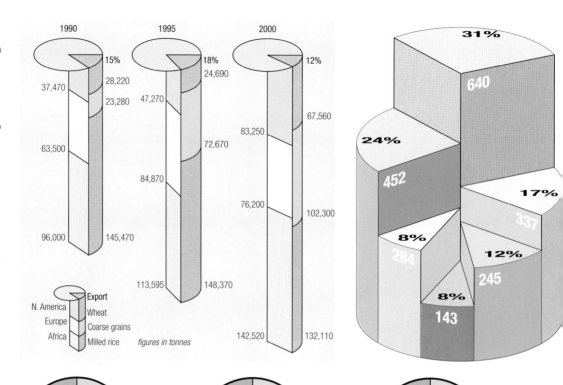

figures in tonnes

Right A third method of combining pie and bar charts allows the pie to appear as a full circle on the face of the bar. The bars vary in length, according to total value, and segments of the bar show an alternative breakdown of the total values.

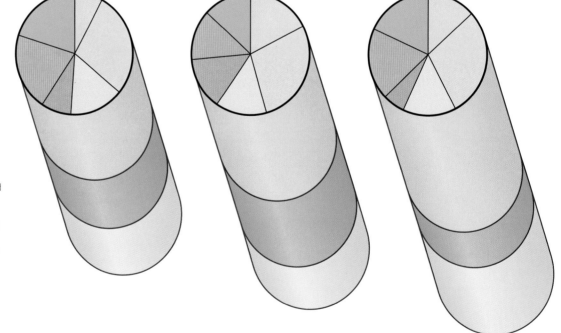

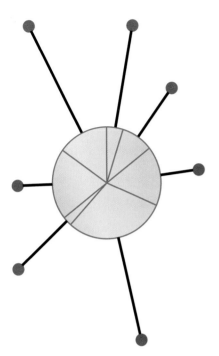

Left You can also add bars to a regular pie by simply using the circumference as a base. By keeping the bars narrow and adding a top marker, you can fit bars to even the narrowest segment – but you'll have to rotate the chart to get the best fit in the design area.

Right Here, bigger segments allow bars to be drawn at a broader measure. Using the circular grid device, you can dispense with actual values for the bar totals.

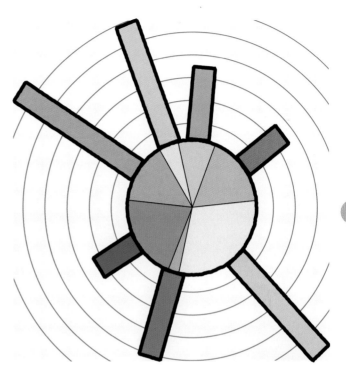

87

Left and right These further variations of the simple pie/bar arrangement place a grid around the bars. By offsetting the grid units, you can make them look like stacks of coins – obviously only used when the bars represent monetary values.

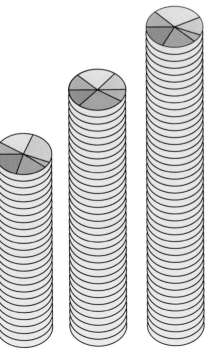

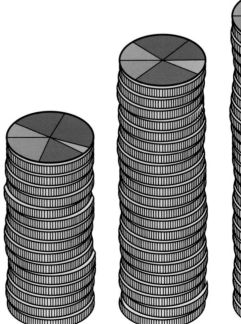

STATISTICAL DIAGRAMS / COMBINED CHARTS 3

Right In this volume chart and divided bar chart combination, the rightmost bar serves as a key to the divisions for all the bars. This allows you to keep the blocks close together and thus get visual value for the volume dimension.

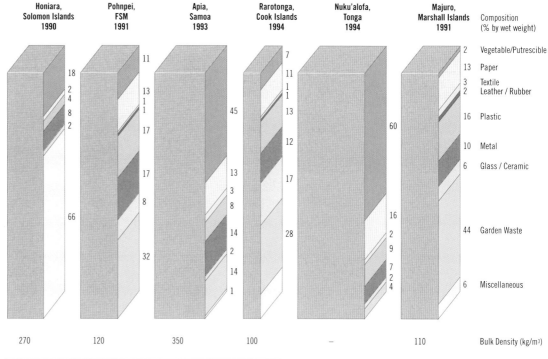

	Honiara, Solomon Islands 1990	Pohnpei, FSM 1991	Apia, Samoa 1993	Rarotonga, Cook Islands 1994	Nuku'alofa, Tonga 1994	Majuro, Marshall Islands 1991	Composition (% by wet weight)
Vegetable/Putrescible	18	11	45	7	60	2	
Paper		13		11		13	
Textile	2	1		1		3	
Leather / Rubber	4	1		1		2	
Plastic	8	17		13		16	
Metal	2			12		10	
Glass / Ceramic		17	13			6	
		8	3	17	16		
Garden Waste	66		8		2	44	
		32	14	28	9		
			2		7		
			14		2		
Miscellaneous			1		4	6	
Bulk Density (kg/m³)	270	120	350	100	–	110	

Right This decorative arrangement of unit-area and variable-area pie charts allows a detailed level of information to be shown. The units are colour-coordinated, with the pie segments enabling the reader to make the association immediately.

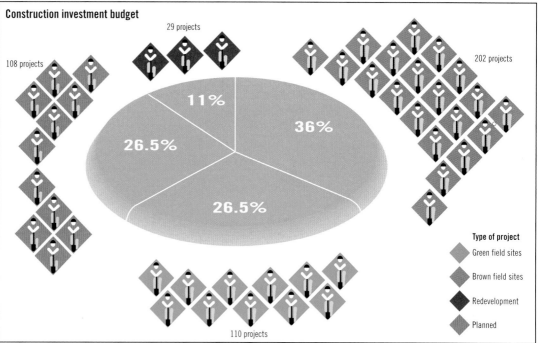

Construction investment budget

108 projects
29 projects
202 projects
11%
36%
26.5%
26.5%
110 projects

Type of project
Green field sites
Brown field sites
Redevelopment
Planned

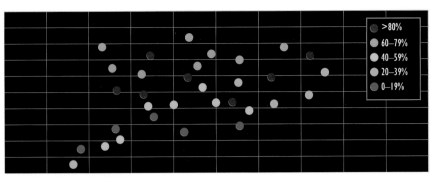

Left Colour use is invaluable as an aid to identification. You can also use it to show values. In this scatter chart, a colour scale lists the values that the colour circles represent. This adds to the information given on the chart's *x* and *y* axes.

Below If colour is a problem, scatter graphs can show the third value as an area chart. The circles are centred on the plot point of the *x*/*y* coordinates, and the size of the circles varies according to the third value.

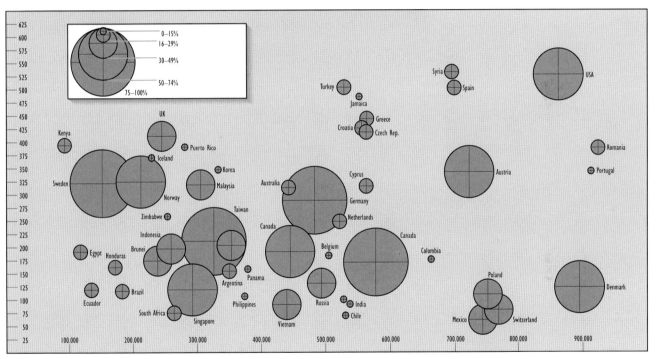

Right Colour is used in this chart, too, to show value. The height of each bar represents average rainfall each week over a year in a particular area, while the colour of the bar shows the average temperature for the same period.

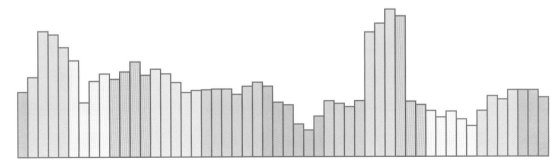

RELATIONAL DIAGRAMS I

Relational diagrams illustrate the physical association or juxtaposition of objects, animate or inanimate, in the physical world. Most commonly these diagrams are maps, plans or charts, in the traditional sense; but they also include signage that tells you, usually in a tabular format, how you relate to particular surroundings – in other words, what your position is.

Plans are probably the oldest kind of graphic device, being an essential tool both for local and more universal navigation. Current navigational techniques make use of artificial-satellite systems and other scientific facilities, yet still depend on visual displays for their functionality. We need to 'picture' where we are.

This section attempts to cover a vast and extremely diverse range of information-design potential in a few pages. It cannot be comprehensive, but should serve as a concise practical guide for the information designer.

90

Right In its simplest form, the relational diagram indicates the direction in which an object lies in relation to the potential user. This caption starts with the word 'right' – which informs you, the reader, that the object described lies in that direction from this starting point.

Left In its simplest form, the relational diagram indicates the direction in which an object lies in relation to the potential user. This caption starts with the word 'left' – which informs you, the reader, that the object described lies in that direction from this starting point.

Right Clear symbols are more visible from a distance than text is, and can be used to clarify text. If they are used as a substitute for written information, their meaning must be unmistakable – particularly where users may travel some distance before realizing their error.

Right Direction signs are basic relational diagrams. As with all information design, they need careful consideration if they are to function efficiently. Take into account where they will be positioned, and the distance from which they are likely to be viewed. A good sign is clearly visible; the content is legible and unambiguous. Direction signs will usually include one or more arrows.

City and Northeast District Hospital

3RD FLOOR

◀ REDWOOD WARD

◀ OAK WARD

◀ MAPLE WARD

◀ BIRCHWOOD WARD

DAY CARE ▶

EYE CLINIC ▶

RADIOLOGY ▶

Right Highway route indicators are a good example of graphic design based on an accepted language. The starting point (the user's position) is always at the base; the current route is drawn on a vertical axis, points upwards, and ends with an arrowhead at the top; subsidiary routes extend from this (usually central) main route and include labels to identify key locations. This basic design can be used in other contexts to show route options. Bear in mind that the majority of users may have little or no local knowledge.

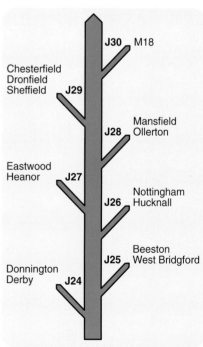

Right Inside public transport vehicles, visible display space is usually horizontal. Line diagrams have to fit to this, and usually simplify the route to a straight line between the terminal points. Legibility is vital, given the often crowded and unsteady state of a mobile vehicle.

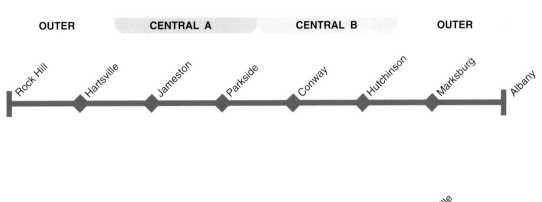

OUTER CENTRAL A CENTRAL B OUTER

Rock Hill · Hartsville · Jameston · Parkside · Conway · Hutchinson · Marksburg · Albany

Right In transport route diagrams, the routes shown will often need to include branch options. With careful design, detailed information can be included – but limit the variety of items shown, and include a clear key.

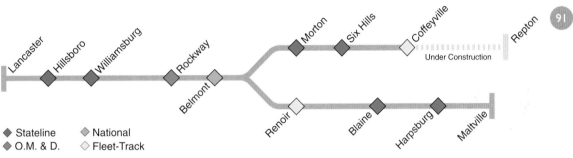

Lancaster · Hillsboro · Williamsburg · Rockway · Belmont · Morton · Six Hills · Coffeyville · Repton
Under Construction
Renoir · Blaine · Harpsburg · Maltville

◆ Stateline   ◆ National
◆ O.M. & D.   ◇ Fleet-Track

Right The simplest maps tend to be limited to locational relationships. The maps here not only show the locational relationship but also indicate travel distance, travel time and travel cost (fares) as comparative lengths of line.

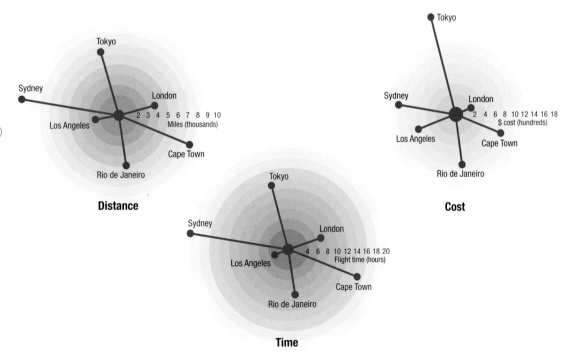

Tokyo
Sydney London
Los Angeles 2 3 4 5 6 7 8 9 10
Miles (thousands)
Rio de Janeiro Cape Town
Distance

Tokyo
Sydney London
2 4 6 8 10 12 14 16 18
$ cost (hundreds)
Los Angeles Cape Town
Rio de Janeiro
Cost

Tokyo
Sydney London
4 6 8 10 12 14 16 18 20
Flight time (hours)
Los Angeles
Cape Town
Rio de Janeiro
Time

RELATIONAL DIAGRAMS 2

Right Plans are the most localized form of mapping. In a basic 2-D floor plan, areas are delineated, the line thickness varying to distinguish wall types. Other features are shown in simple geometric form.

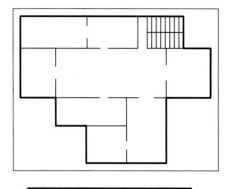

Below Resizing a the vertical scale of a 3-D floor plan to varying degrees gives a good perspective effect. You'll need to adjust the wall depths.

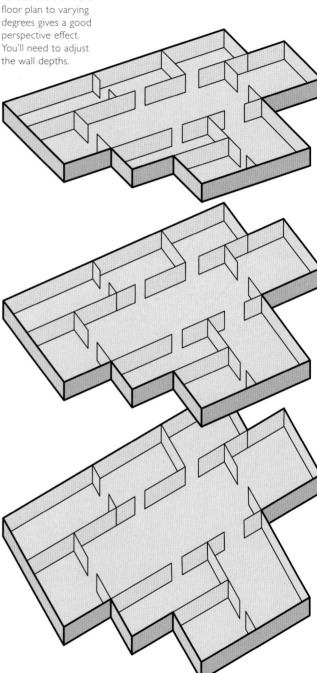

92

Right Adding height to the wall introduces the third dimension and transforms the plan, giving the reader a stronger visual idea of the structure. Additional items, such as stairs, become a little trickier to draw, but the overall result is more solid.

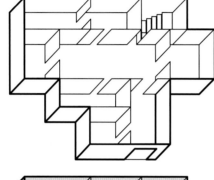

Right You can further enhance the 3-D effect by adding tones of colour to the walls and floor areas.

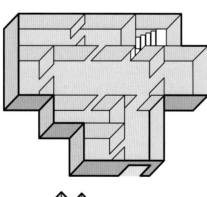

Right Working to a non-perspective projection makes it easier to do the drawing with 2-D software. You could use any angle, but established projections such as the isometric or oblique projections (see page 60) are most effective.

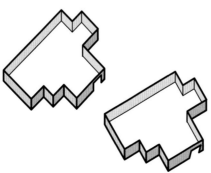

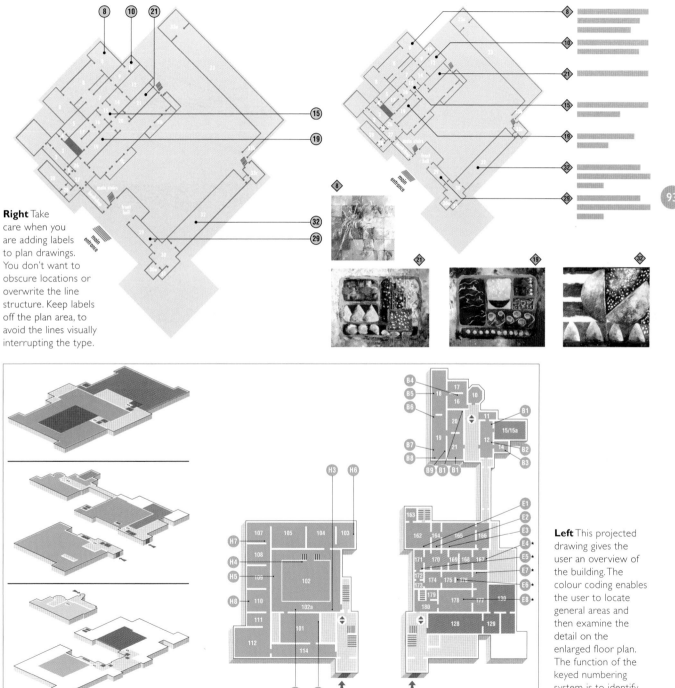

Right Take care when you are adding labels to plan drawings. You don't want to obscure locations or overwrite the line structure. Keep labels off the plan area, to avoid the lines visually interrupting the type.

Left This projected drawing gives the user an overview of the building. The colour coding enables the user to locate general areas and then examine the detail on the enlarged floor plan. The function of the keyed numbering system is to identify specific locations.

RELATIONAL DIAGRAMS 3

Right Local relational diagrams in the form of street maps need careful preparation. Streets have to be drawn so that they relate to each other, but can be simplified to a considerable extent. The examples here range from the realistic to a stylized geometric treatment.

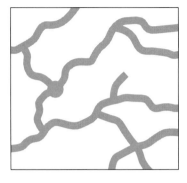
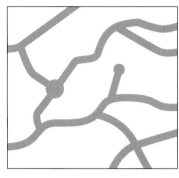
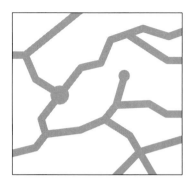

Right How you draw the actual street lines is a matter of style. A tint base suggests a built-up area with white streets. Draw a wide black line and superimpose a cloned but thinner white line to obtain parallel road edges.

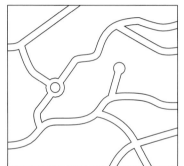
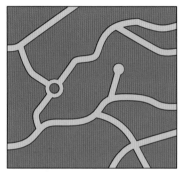

Right Labels usually follow the line of the street. Drawing software allows you to attach type to invisible paths formed by cloned street lines. Don't rotate type through more than 90°.

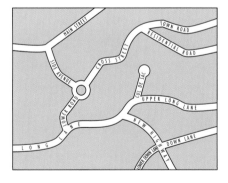

Right All too often you'll find that street names are longer than the streets on the plan. Leader lines and abbreviations will help. You'll also find that several names appear along one stretch of street. If the names are close together, try using a separating device.

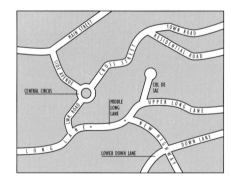

Right A matrix of grid lines dividing a large plan into smaller units, combined with a street index, aids the reader in locating streets on complex maps. Make sure the grid lines are clear, but don't interrupt or confuse the lines of the streets.

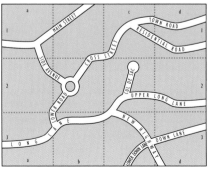

Right It is generally accepted that maps and plans are drawn with north at the top. If your map is rotated, you should indicate this by using a grid or a north pointer. Most maps will need some form of scale of distance (remember to include the scale in any change of size).

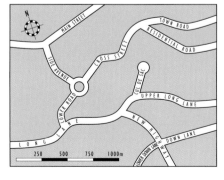

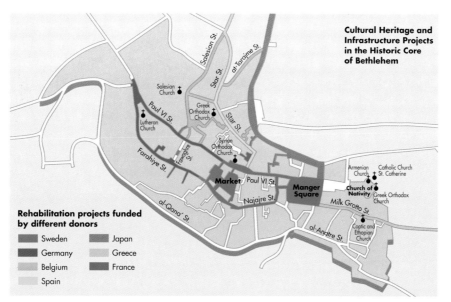

Cultural Heritage and Infrastructure Projects in the Historic Core of Bethlehem

Salesian St.
Star St.
at-Tarajme St.
Salesian Church
Paul VI St.
Greek Orthodox Church
Lutheran Church
Star St.
Farahiye St.
Fawaghre St.
Syrian Orthodox Church
Armenian Church
Catholic Church St. Catherine
Market
Paul VI St.
Church of Nativity
Greek Orthodox Church
Manger Square
Najaire St.
Milk Grotto St.
al-Qana' St.
al-Anatre St.
Coptic and Ethopian Church

Rehabilitation projects funded by different donors

- Sweden
- Germany
- Belgium
- Spain
- Japan
- Greece
- France

Left Once you have drawn the basic street plan you can incorporate other information. Colour will obviously aid visual differentiation of areas or streets.

95

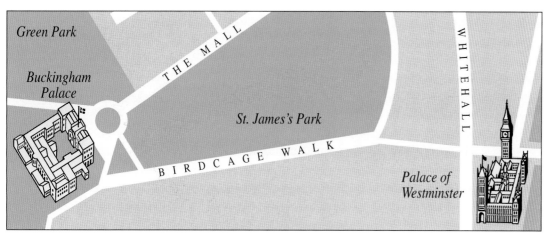

Green Park

Buckingham Palace

THE MALL

WHITEHALL

St. James's Park

BIRDCAGE WALK

Palace of Westminster

Left Pictograms are a good way of adding interest to a simple street plan.

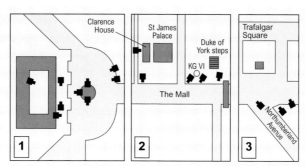

1 Buckingham Palace and Canada gate.
7 cameras (9 positions)

2 Clarence House and the Mall.
5 Cameras

3 Trafalgar Square.
3 Cameras

Clarence House

St James Palace

Duke of York steps

Trafalgar Square

KG VI

The Mall

Northumberland Avenue

1   **2**   **3**

Left Keep the representation of streets as simple as possible on diagrams such as this which serve only as a background to locate particular activities. Buildings, too, are shown in simple geometric form.

RELATIONAL DIAGRAMS 4

Right Sometimes you will need to show information at local area level together with a wider over-view. Rather than attempting to show the detail on a single map, you could use two or more maps with different scales. At each level, the next stage map is shown as an inset area. Varying the size of the map, with the biggest map being the main focus, creates an interesting arrangement. For a more dynamic effect, you can rotate and angle the maps.

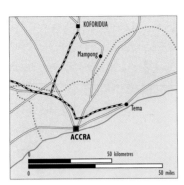

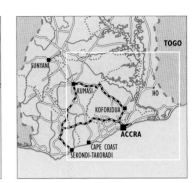

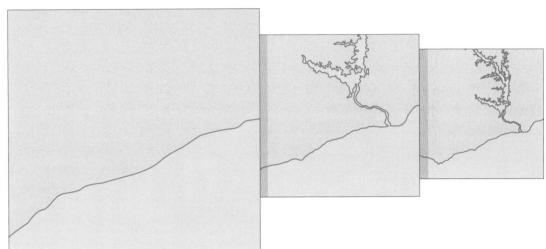

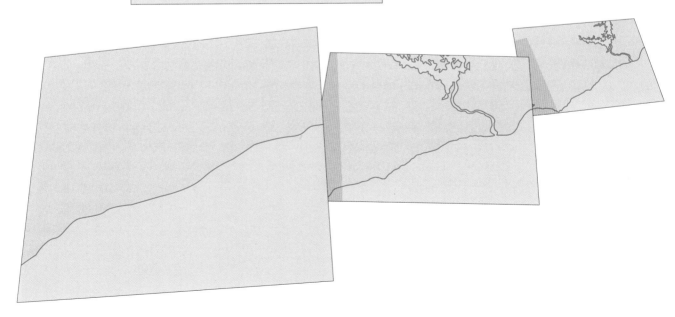

Right Stylize map detail by plotting it along a geometric grid. In the example, all lines change direction in multiples of 45°. This departs from the actual geographical form, but creates a striking graphic effect.

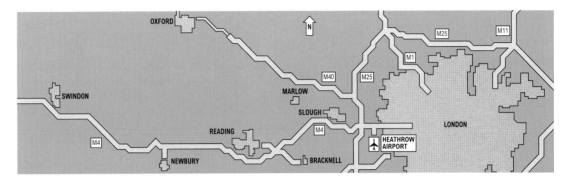

Below Use of strong colour produces eye-catching maps. Here, the map detail has been limited to items of relevance to the subject. The inset map locates the large-scale map in a broader geographical context.

Right In this map, too, the content is restricted to relevant subject matter. The underlying tonal base, indicating topographic structure, enhances the simple treatment of the other items.

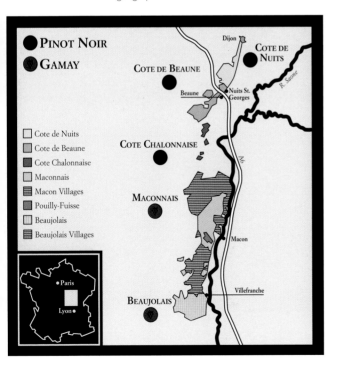

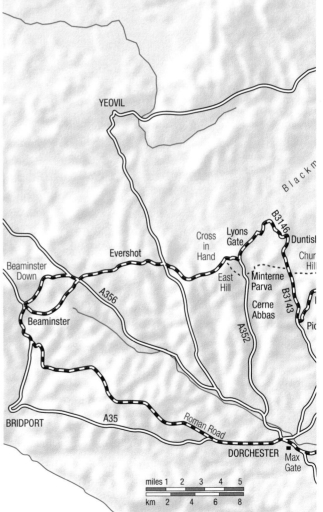

RELATIONAL DIAGRAMS 5

Right and below
Maps at the global level present particular problems for the information designer. The world is (almost) spherical. At lower levels, areas can be assumed to be flat and evenly scaled in either dimension. To deal with mapping at this level requires a basic understanding of geography in order to avoid the pitfalls.

Right and below
To draw usable maps of the 3-D globe in two dimensions, cartographers have devised all sorts of projections. Each one is a compromise, distorting land areas to a greater or lesser extent. So designers have to select the most appropriate one for their needs.

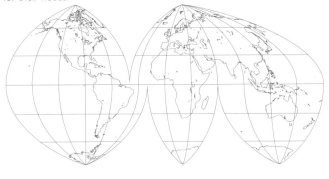

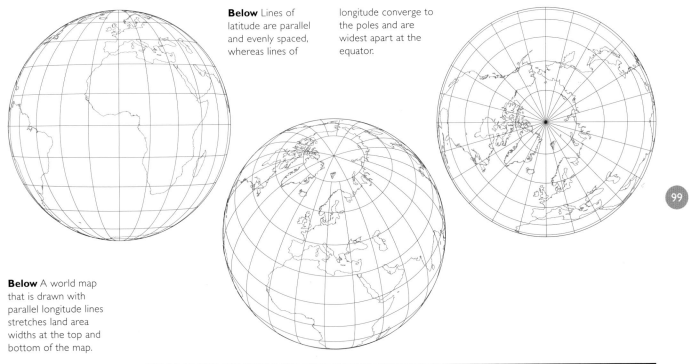

Below Lines of latitude are parallel and evenly spaced, whereas lines of longitude converge to the poles and are widest apart at the equator.

Below A world map that is drawn with parallel longitude lines stretches land area widths at the top and bottom of the map.

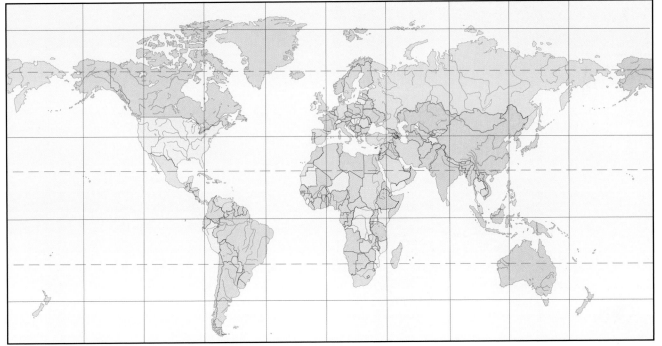

RELATIONAL DIAGRAMS 6

Right Because reality has to be edited to work on the 2-D page, all mapping is a compromise. You will need to decide the level of editing to suit the use. If too much detail is drawn, it will fill in and look ugly, so bear in mind the size at which it will be reproduced.

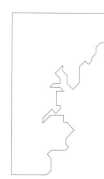

Right This selection of examples shows some of the many styles you can use for coastlines. Unlike the cartographer, the information designer can experiment to obtain striking results. Digital technology facilitates such experimentation.

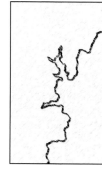

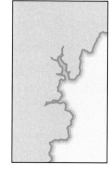

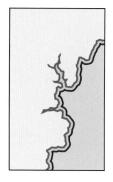

Right Lakes, rivers and other waterways can be treated in a variety of ways, although readers tend to associate blue with water.

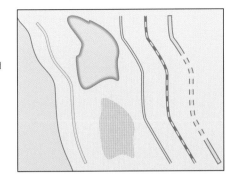

Right You can show hills and mountains by symbols. Trees, too, can be represented by icons. Larger areas of woodland can be shown by colour.

Right Regional and national boundaries can be drawn as lines. You might need to use a different style to show disputed national borders, of which there are many. Alternatively, just distinguish areas by colour fill.

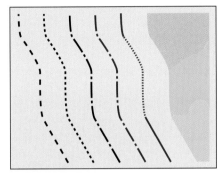

Right Various line styles (both treatment and colour) are used to represent roads and other transport links. Be consistent in line thickness for types of road. Draw parallel lines digitally by superimposing a duplicated line over the original and making it narrower.

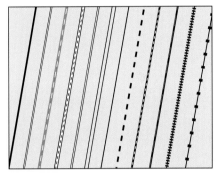

Right Town locations may be represented by dots on small-scale maps, or by patches of tint simulating the urban shape on large-scale ones. You can vary the size or style of dot to indicate the size or significance of the town.

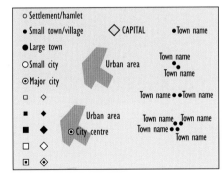

o Settlement/hamlet
● Small town/village
● Large town
○ Small city
◉ Major city
◇ CAPITAL
Urban area
●Town name
Town name
Town name
Town name ● ●Town name
Urban area
◉ City centre
Town name ●● Town name
Town name, Town name
Town name

Right You can use icons to locate other significant points. There are fonts that include wide collections of appropriate symbols, or you can devise your own. Just make sure that they will read at the size at which they are to be reproduced.

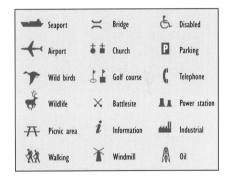

Seaport	Bridge	Disabled
Airport	Church	Parking
Wild birds	Golf course	Telephone
Wildlife	Battlesite	Power station
Picnic area	i Information	Industrial
Walking	Windmill	Oil

Right A compass point is useful, particularly if the map is not north-oriented. Most maps benefit from the inclusion of a scale. This needs to be long enough for the reader to make sensible use of it, but not so long as to intrude on the map detail.

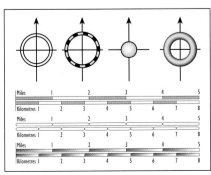

Miles	1	2	3	4	5
Kilometres	1	2	3	4	5	6	7	8
Miles	1	2	3	4	5
Kilometres	1	2	3	4	5	6	7	8
Miles	1	2	3	4	5
Kilometres	1	2	3	4	5	6	7	8

Right You may want to finish your map with a simple single line frame, or you could give it a little extra style with a more elaborate edge. The frame can also be used as a clipping path, to tidy up roads and other lines that overlap the edge.

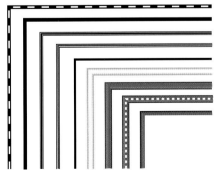

RELATIONAL DIAGRAMS 7

Right The same
map, different effects
– all achieved by
means of digital
manipulation.

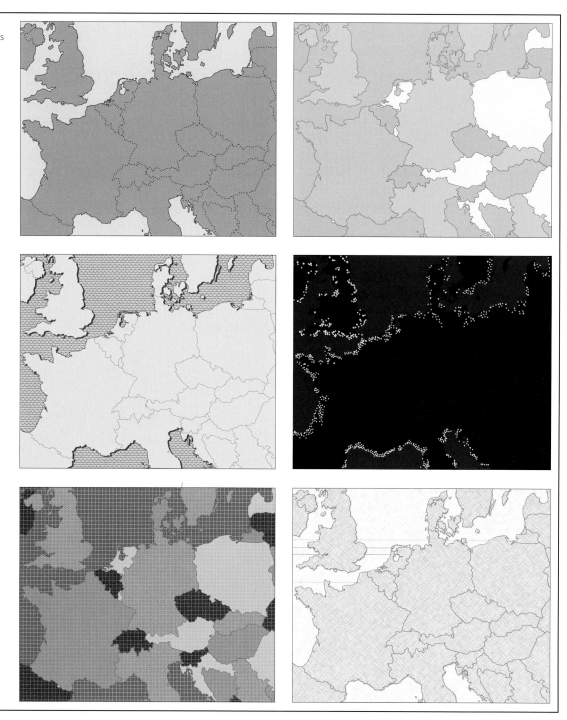

Right You can create striking images with perspective effects generated by digital manipulation.

Right Less dramatic, but equally effective, is this 'antique' styling. The lines are made uneven and a hand-drawn font is used for type labels in order to achieve the effect.

RELATIONAL DIAGRAMS 8

Contours show height variations. This cross-section through a mountain has a vertical scale of heights. If lines are drawn on the surface, where the scale touches you can see the linear pattern of contours.

Right Using the contour lines, you can show different heights by colour. A common convention in atlases, this can be used to great effect by the information designer.

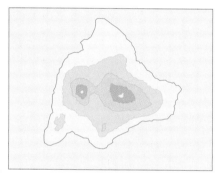

Right Also a fairly standard convention in atlases is the addition of shading. This usually assumes a light-source at top left. It needs to be subtle if labels and other graphic matter are to be legible.

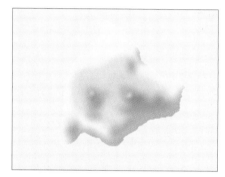

Right This unusual but effective treatment is obtained by adding a soft shadow to the underside of each contour level.

Right Hachures were used extensively to show topographic information in atlases of hand-drawn maps. Although effective, they are very time-consuming to create, even digitally.

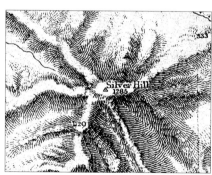

Right You can render realistic terrain by using 'digital elevation models' (DEMs). Data for these is freely available from various sources on the Web. The converted data provides a greyscale image – white is high, black low.

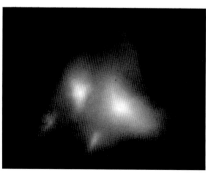

Right Using the 'Lighting Effects' feature of an image-editing application such as Adobe Photoshop, you can use a greyscale DEM as a 'bump map' to render a realistic terrain view.

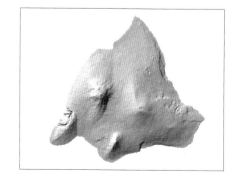

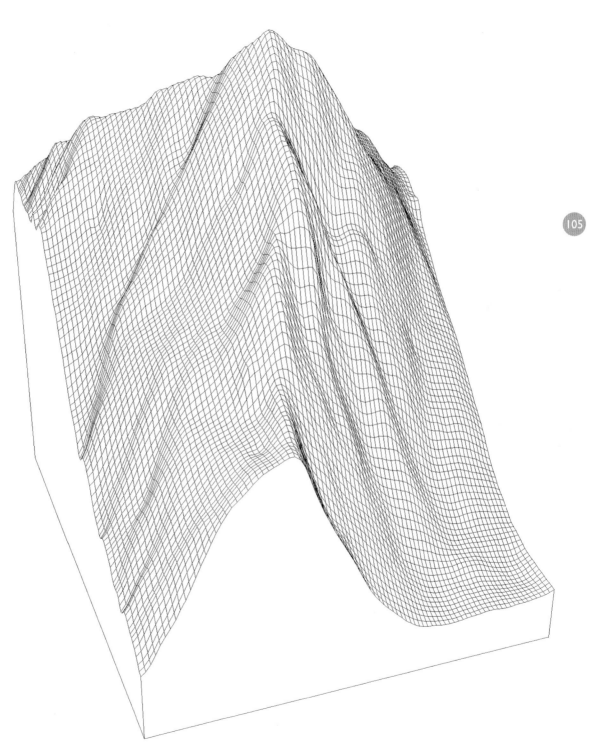

Right Check out the graphic capabilities of your spreadsheet program. Some are able to display and manipulate contour information derived from X-Y-Z coordinates, such as in this image of Mount Everest. Once compiled, the image can then be angled and rotated as you wish.

RELATIONAL DIAGRAMS 9

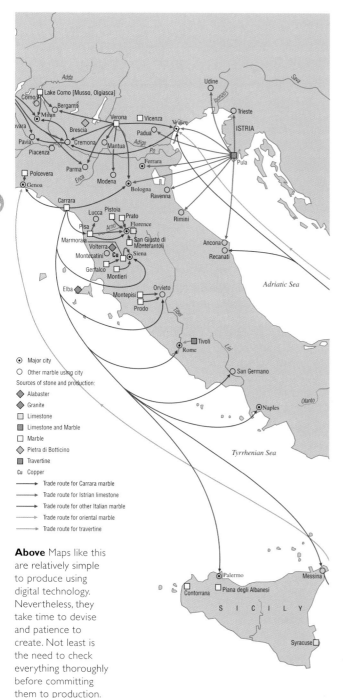

Map legend (left):

- ⊙ Major city
- ○ Other marble using city

Sources of stone and production:

- ◆ Alabaster
- ◆ Granite
- ☐ Limestone
- ▨ Limestone and Marble
- ☐ Marble
- ◇ Pietra di Botticino
- ▨ Travertine
- Cu Copper

— Trade route for Carrara marble
— Trade route for Istrian limestone
— Trade route for other Italian marble
— Trade route for oriental marble
— Trade route for travertine

Above Maps like this are relatively simple to produce using digital technology. Nevertheless, they take time to devise and patience to create. Not least is the need to check everything thoroughly before committing them to production.

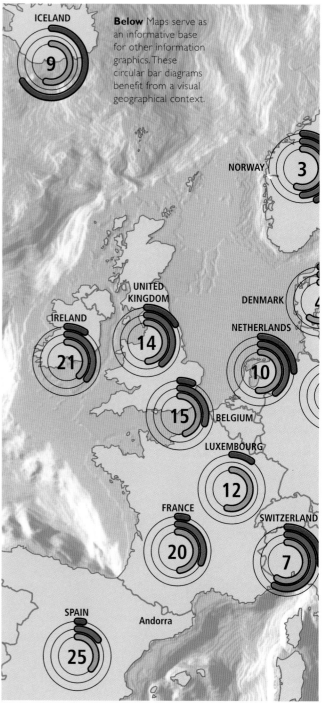

Below Maps serve as an informative base for other information graphics. These circular bar diagrams benefit from a visual geographical context.

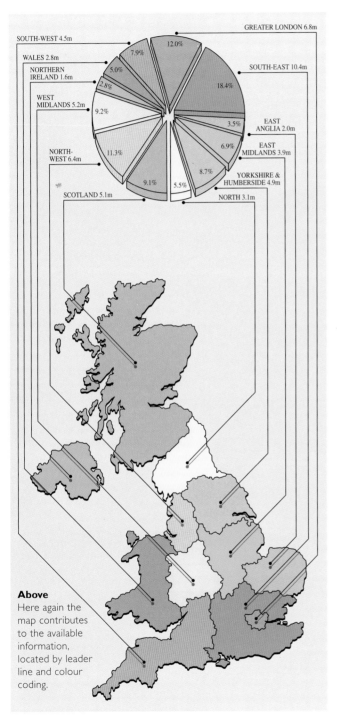

Above
Here again the map contributes to the available information, located by leader line and colour coding.

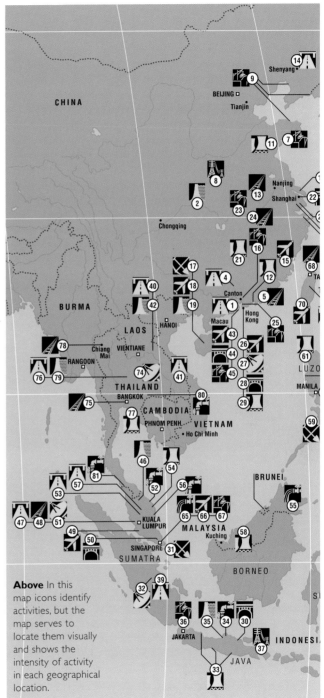

Above In this map icons identify activities, but the map serves to locate them visually and shows the intensity of activity in each geographical location.

ORGANIZATIONAL DIAGRAMS 1

Whereas relational charts are used to illustrate a general pattern of links, organizational diagrams show specific interrelationships between items. These items need not be physical entities, but can be abstract concepts or activities.

Organizational diagrams have several uses. They appear commonly in the form of a 'family tree' – which is used not only to map family ancestry but also to plot the often complex historical process of business mergers and acquisitions. The 'organogram' shows the inherent structure of an organization.

Other types of organizational diagram include the flow chart (used to illustrate one or more flow patterns) and the diagnostic chart, a variation of the flow chart that shows alternative 'routes' decided by the answers to question embedded in the diagram.

108

Right Organizational diagrams link units of information. The units are usually shown contained in boxes, with some form of line connecting selected units. Weight of line and style of box can be varied in order to indicate different relationships.

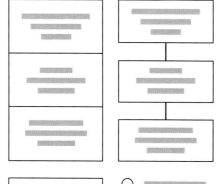
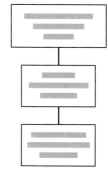
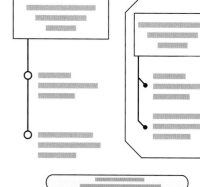

Right You can vary the style of box to show differing status within the structure, and use different line weights, colours and styles to indicate different types of relationship.

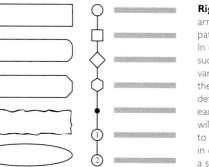

Right In flow charts, arrows indicate the path to be followed. In diagnostic charts, such as this, the route varies according to the condition that is determined following each question. You will frequently need to make several drafts in order to achieve a satisfactory design.

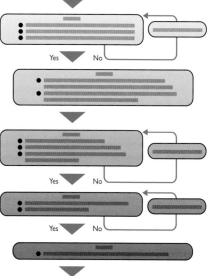

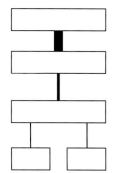
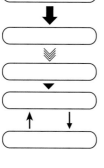
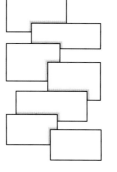
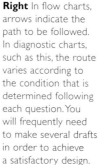

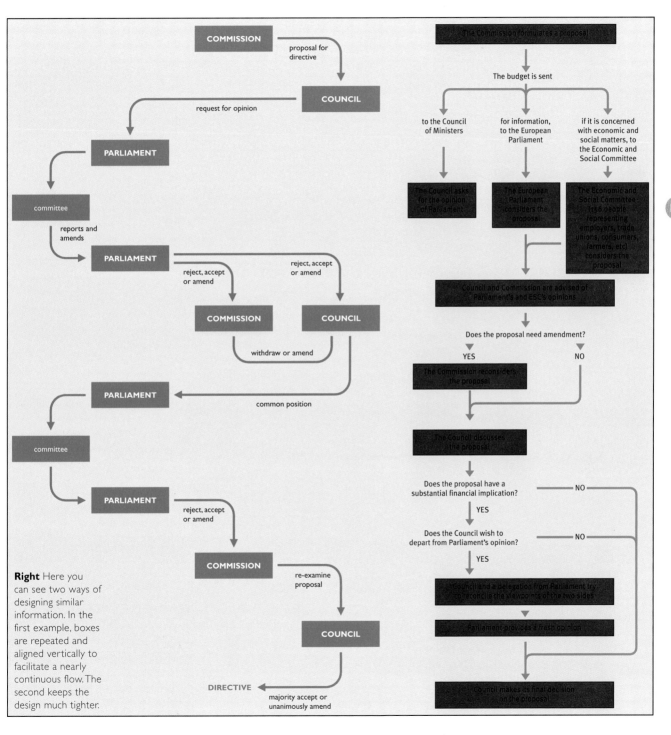

COMMISSION

proposal for
directive

COUNCIL

request for opinion

PARLIAMENT

committee

reports and
amends

PARLIAMENT

reject, accept
or amend

reject, accept
or amend

COMMISSION

COUNCIL

withdraw or amend

PARLIAMENT

common position

committee

PARLIAMENT

reject, accept
or amend

COMMISSION

re-examine
proposal

COUNCIL

DIRECTIVE

majority accept or
unanimously amend

Right Here you
can see two ways of
designing similar
information. In the
first example, boxes
are repeated and
aligned vertically to
facilitate a nearly
continuous flow. The
second keeps the
design much tighter.

The Commission formulates a proposal

The budget is sent

to the Council
of Ministers

for information,
to the European
Parliament

if it is concerned
with economic and
social matters, to
the Economic and
Social Committee

The Council asks
for the opinion
of Parliament

The European
Parliament
considers the
proposal

The Economic and
Social Committee
(156 people
representing
employers, trade
unions, consumers,
farmers, etc)
considers the
proposal

Council and Commission are advised of
Parliament's and ESC's opinions

Does the proposal need amendment?

YES

NO

The Commission reconsiders
the proposal

The Council discusses
the proposal

Does the proposal have a
substantial financial implication?

NO

YES

Does the Council wish to
depart from Parliament's opinion?

NO

YES

Council and a delegation from Parliament try
to reconcile the viewpoints of the two sides

Parliament provides a fresh opinion

Council makes its final decision
on the proposal

ORGANIZATIONAL DIAGRAMS 2

Right The 'family tree' is used to show how corporations have grown by mergers and takeovers. In this example, the boxes are positioned in a vertical chronological order – which, given the space constraints, can only be relative.

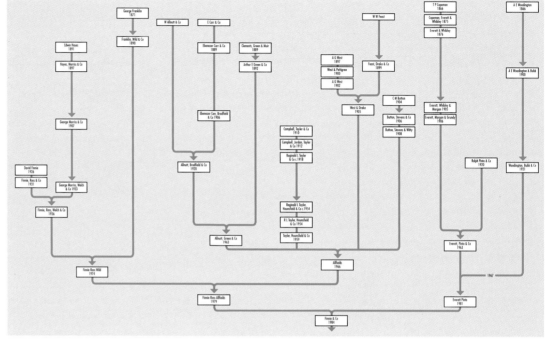

Right This family tree is an organizational chart turned through 90° and stripped of the usual boxes. With so many names to fit into a tight area, the names within each generation have been made into vertical lists, which take up less space. Leaving off the boxes also saves space. Emboldening the names of direct descendants helps the reader to 'navigate' through the chart.

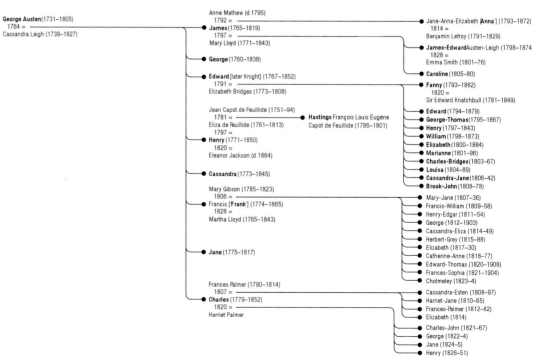

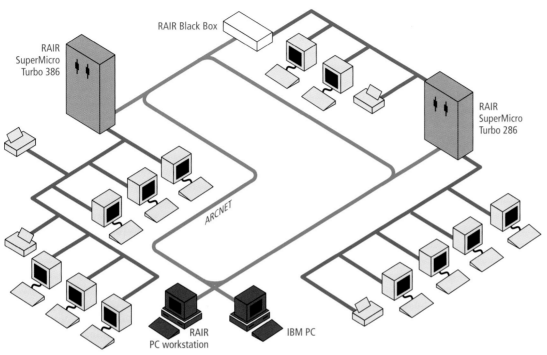

RAIR Black Box

RAIR
SuperMicro
Turbo 386

RAIR
SuperMicro
Turbo 286

ARCNET

RAIR
PC workstation

IBM PC

Left Icons and
pictograms not only
add visual appeal but
are also employed to
aid understanding of
organizational charts.

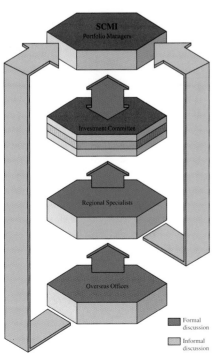

SCMI
Portfolio Managers

Investment Committee

Regional Specialists

Overseas Offices

Formal
discussion

Informal
discussion

Left Unusual shapes
shown in 3-D give
interest to what
would otherwise be a
very basic chart.

Right This diagram,
showing values of
imports and exports,
uses the width of the
arrows to display the
quantities.

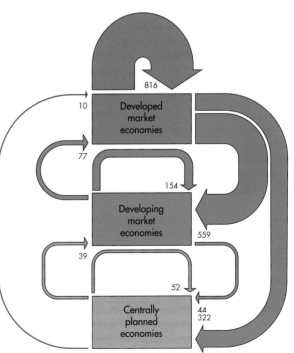

816

10

Developed
market
economies

77

154

Developing
market
economies

559

39

52

Centrally
planned
economies

44
322

ORGANIZATIONAL DIAGRAMS 3

Right In some cases you might want to avoid a hierarchical structure for an organizational chart. This 'organogram' sites the head at the centre, with the lines of management communication radiating out to the subsidiary divisions.

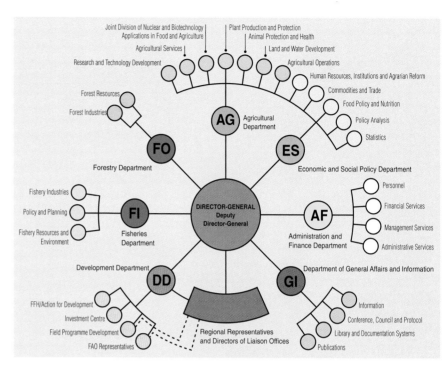

Joint Division of Nuclear and Biotechnology
Applications in Food and Agriculture

Agricultural Services

Research and Technology Development

Plant Production and Protection
Animal Protection and Health

Land and Water Development

Agricultural Operations

Human Resources, Institutions and Agrarian Reform

Commodities and Trade

Food Policy and Nutrition

Policy Analysis

Statistics

Forest Resources

Forest Industries

AG Agricultural Department

FO

ES

Forestry Department

Economic and Social Policy Department

Fishery Industries

Policy and Planning

Fishery Resources and Environment

Personnel

Financial Services

Management Services

Administrative Services

FI

DIRECTOR-GENERAL
Deputy
Director-General

AF

Fisheries Department

Administration and Finance Department

Development Department

Department of General Affairs and Information

DD

GI

FFH/Action for Development

Investment Centre

Field Programme Development

FAO Representatives

Information

Conference, Council and Protocol

Library and Documentation Systems

Publications

Regional Representatives
and Directors of Liaison Offices

Right Add interest to an organizational diagram by drawing it at an angle.

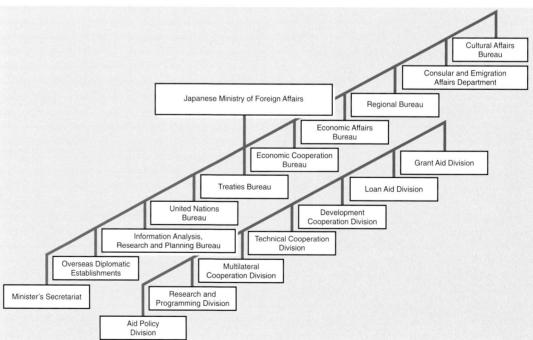

Japanese Ministry of Foreign Affairs

Cultural Affairs Bureau

Consular and Emigration Affairs Department

Regional Bureau

Economic Affairs Bureau

Economic Cooperation Bureau

Grant Aid Division

Treaties Bureau

Loan Aid Division

United Nations Bureau

Development Cooperation Division

Information Analysis, Research and Planning Bureau

Technical Cooperation Division

Overseas Diplomatic Establishments

Multilateral Cooperation Division

Minister's Secretariat

Research and Programming Division

Aid Policy Division

Right You can reduce some flow charts to a relatively simple structure of arrows and text – which excludes unnecessary visual clutter.

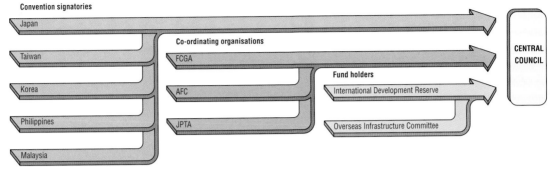

Below This cyclical flow diagram shows the physical relationship of the elements. It is designed to serve both as an aid to understanding and as a decorative style treatment.

Right Incorporating icons or pictograms in a chart not only adds interest but can be an effective way of saving space, especially when long scientific names would otherwise have to be included – and repeated.

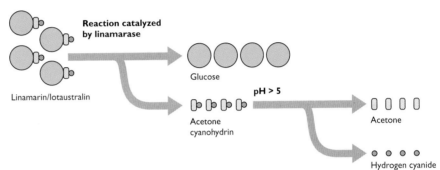

113

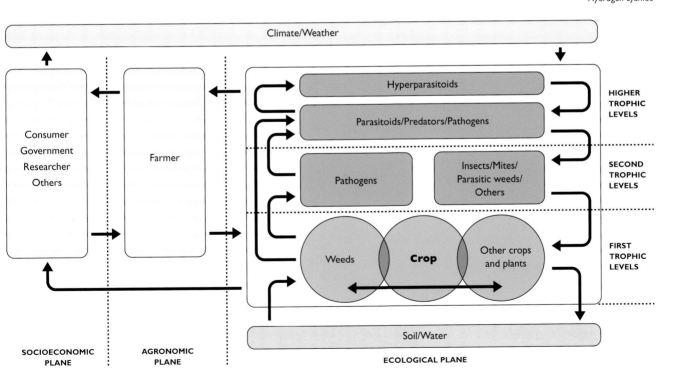

TIME DIAGRAMS 1

Organizing the presentation of time in tables or charts has its own particular problems. However, as with any other kind of information graphics, intelligent and sensitive design control will produce an invaluable tool for the reader, allowing the uninterrupted assimilation of information.

Apart from straightforward calendars or almanacs, time charts will need to include variable, and often substantial, amounts of type matter. This may well create problems when a timescale of equal division is required. Even when divisions can be varied, events involving substantial wordage are unlikely to occur conveniently side-by-side. In such cases, skilful design is needed to ensure that the facts are displayed clearly in an easily understood chronological matrix.

114

Right There are special types of table that have their own design considerations. A TV schedule can be made to function as an easy-to-read listing by arranging the column entries chronologically. The graphic clock device enables the user to find desired viewing times by quickly scanning down the left-hand column, while horizontal rules group programme start times within each hourly period.

Transport timetables may include routes that have non-stop variations. Usually departure and arrival times are in chronological order, but this means some time lists will cross over. The convention is to break the list that is overlapped. A simple graphic enhancement like this soft shadow appears to 'lift' the overlapping, express route. This has the joint effect of drawing attention to the limited-stop service and making the link to the split list easier to visualize. A more elaborate graphic treatment shows the route as linked circles, but leaves open circles for the through route. The line for the slower service remains continuous, and so simpler to comprehend.

Moston	09.06	09.36	09.21	
Kirbeck	09.11	09.41		
Perton	09.21	09.51		
Suskwell	09.27	09.57		
Ormby	09.31	10.01 ▶		▶ 10.01
Newdyke	09.34		09.50	10.04
Kilbyton	09.38			10.08
Irborough	09.41			10.11
Ulanwood	09.46			10.16
Chitern	09.54			10.24
Klackford	10.10		09.58	10.40

Moston	09.06	09.36	09.21	
Kirbeck	09.11	09.41		
Perton	09.21	09.51		
Suskwell	09.27	09.57		
Ormby	09.31	10.01 ▲		
Newdyke	09.34		09.50	▲10.04
Kilbyton	09.38			10.08
Irborough	09.41			10.11
Ulanwood	09.46			10.16
Chitern	09.54			10.24
Klackford	10.10		09.58	10.40

Moston	● 09.06	● 09.36	● 09.21	
Kirbeck	● 09.11	● 09.41	○	
Perton	● 09.21	● 09.51	○	
Suskwell	● 09.27	● 09.57	○	
Ormby	● 09.31	● 10.01	○	
Newdyke	● 09.34		● 09.50	●10.04
Kilbyton	● 09.38			●10.08
Irborough	● 09.41			●10.11
Ulanwood	● 09.46			●10.16
Chitern	● 09.54			●10.24
Klackford	● 10.10		● 09.58	●10.40

Right Apply a little imagination and a few simple graphics – and what might have been a boring listing of information can be transformed into a useful and interesting design.

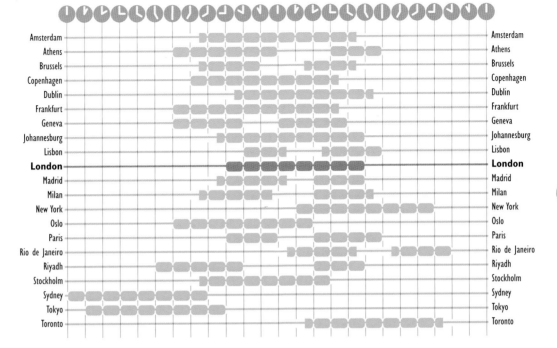

Amsterdam
Athens
Brussels
Copenhagen
Dublin
Frankfurt
Geneva
Johannesburg
Lisbon
London
Madrid
Milan
New York
Oslo
Paris
Rio de Janeiro
Riyadh
Stockholm
Sydney
Tokyo
Toronto

Amsterdam
Athens
Brussels
Copenhagen
Dublin
Frankfurt
Geneva
Johannesburg
Lisbon
London
Madrid
Milan
New York
Oslo
Paris
Rio de Janeiro
Riyadh
Stockholm
Sydney
Tokyo
Toronto

115

Right You can even turn a simple but repetitive list of dates into an attractive visual image. Again, all you need is a little imagination and a few simple graphic tricks.

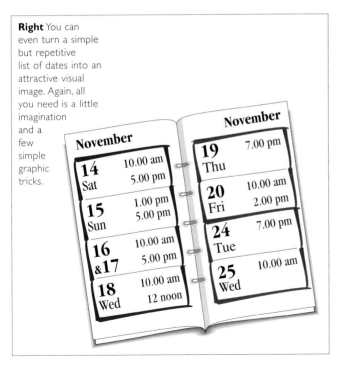

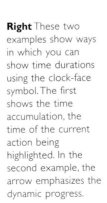

November	
14 Sat	10.00 am / 5.00 pm
15 Sun	1.00 pm / 5.00 pm
16 &**17**	10.00 am / 5.00 pm
18 Wed	10.00 am / 12 noon

November	
19 Thu	7.00 pm
20 Fri	10.00 am / 2.00 pm
24 Tue	7.00 pm
25 Wed	10.00 am

Right These two examples show ways in which you can show time durations using the clock-face symbol. The first shows the time accumulation, the time of the current action being highlighted. In the second example, the arrow emphasizes the dynamic progress.

TIME DIAGRAMS 2

Right Cyclical time patterns can often be plotted as circular charts. This chart combines several related elements of information. The repetitious nature of the process is made particularly clear by the continuity of the circle.

Menstrual period

Pituitary gland produces follicle stimulating hormone (FSH)

Pituitary gland produces luteinizing hormone (LH)

Below Another
useful visual device
that can be employed
in time-chart design
is the calendar block.
This creates a focus
for the particular
date, or it can be the
commencement point
for a month or year.

Below You can plot
timelines vertically or
horizontally. Shown
below is an example
of a vertical time
chart in which the
entries are also offset
along a horizontal
relative timescale.

Below If entries
vary in proximity and
line length, use leader
lines to relate the
points to an exact
position along the
timescale. This should
give you room to
accommodate longer
entries with enough
separation to ensure
legibility.

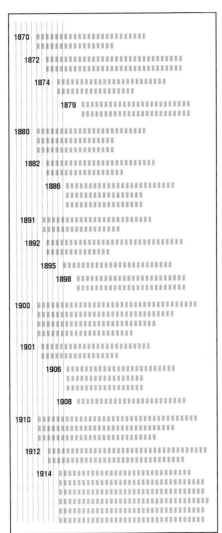

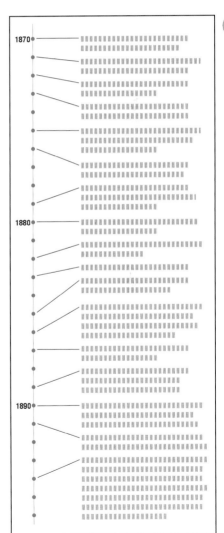

TIME DIAGRAMS 3

Right As time is usually perceived as a linear progression, timelines are often plotted to read from left to right. This makes it difficult to fit in long entries. Offset them vertically, and use a bullet device to align the points visually to the timescale.

Right Alternatively, you can run the scale through the middle of the chart area and use angled type to accommodate the line entries.

Right Timelines may need to cover long durations. Simple perspective enables you to compress the timeline. It also allows more space for fitting text elements relating to recent events.

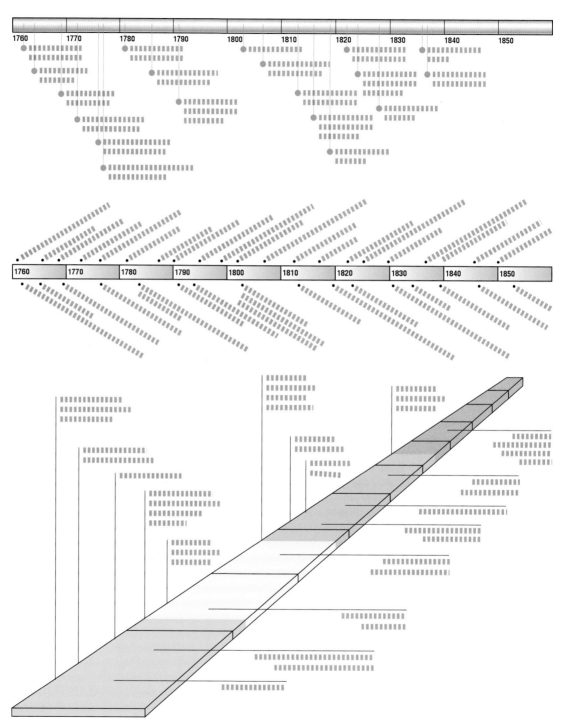

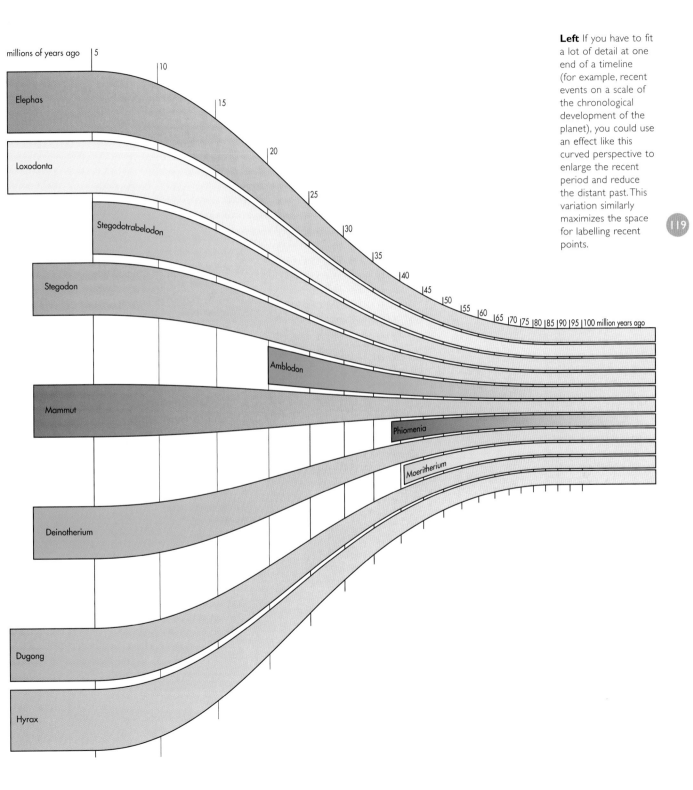

millions of years ago |5

|10

|15

|20

|25

|30

|35

|40

|45

|50

|55

|60

|65 |70 |75 |80 |85 |90 |95 |100 million years ago

Elephas

Loxodonta

Stegodotrabelodon

Stegodon

Amblodon

Mammut

Phiomenia

Moeritherium

Deinotherium

Dugong

Hyrax

Left If you have to fit a lot of detail at one end of a timeline (for example, recent events on a scale of the chronological development of the planet), you could use an effect like this curved perspective to enlarge the recent period and reduce the distant past. This variation similarly maximizes the space for labelling recent points.

PAGE-LAYOUT SOFTWARE / 1

Within just a few years a vast array of tools has been put at the disposal of the information designer. Previously, you could spend hours carefully plotting a chart only to have to abandon a rejected design or take in an adjustment to data that required the reworking of the drawing. Or even worse, you might have to make a substantial amendment to a design that had gone all the way to final artwork. These days you can spend some time experimenting, you can change fonts or colours with ease, and you can resize or reformat an image easily within the schedule. Life could not be easier – except that clients know that these changes can be made fairly simply and quickly, and so seem inclined to make late changes just for the sake of it.

The specialist tools in the designer's toolbox are indeed powerful. With relatively inexpensive 2-D and 3-D vector drawing software (*see page 130*) you can achieve striking results. And these can be enhanced with raster painting programs (*see page 134*) to produce realistic rendering of textures and create extremely convincing images.

Yet even without specialist drawing software, it is possible to generate surprisingly good diagrams. Page-layout programs now have drawing facilities built in. Up-to-date spreadsheet programs have diagram-creation facilities, although the design controls in these can be frustratingly limited. Experiment with word-processing programs. If they can draw lines and boxes to a specified dimension, you can produce bar charts, area charts and flow charts. With a little imagination and some patience, you can produce highly professional-looking graphics with some very basic tools (see the map created within a page-layout program, for example).

The following pages show a number of examples and techniques, using standard professional packages readily available to the digital-diagram designer.

Below Varying the shade and position of rules on a tinted base gives the illusion of embossed or engraved effects. Lines are simply abutted by duplicating and moving by the line width. Adding lines at 90° produces boxes with the same effects.

Below Using a
Bézier drawing tool,
you can draw a filled
graph. Make a thin
round-cornered box
and adjust its angle,
by trial and error,
to match a segment
of the graph line.
Duplicate the box
and adjust the angle
to create the graph
line as a broad rule.

You can then
duplicate and offset
the line to create
a shadow. You can
also create a shadow
for the grid.

Below To produce a
single line graph, you'll
have to draw each
of the line segments
individually, and the
ends probably won't
abut nicely. Use a
small point marker to
disguise this. Or you
can make a feature
of the point marker.
These balls are simply
circular boxes with

a graduated fill and
a small highlight.
Alternatively, having
drawn the Bézier
outline, you could
create a series of
parallel vertical lines
and extend these to
align their tops with
the graph line.

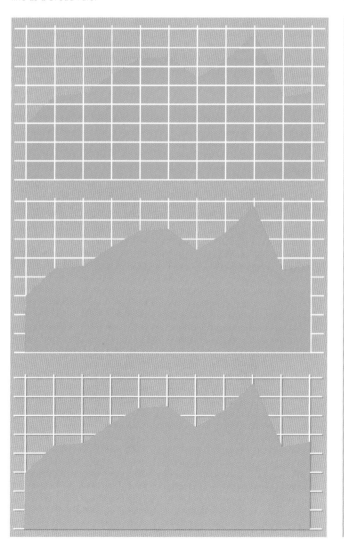

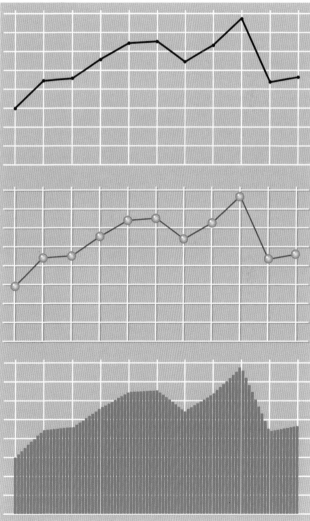

PAGE-LAYOUT SOFTWARE / 2

Below Simple filled boxes created with a page-layout program make basic bar charts. Use graduated tints of colours to produce more striking effects. To make shadow effects, duplicate the bars, then change the colours to grey and position them behind.

Below Each of these metallic-looking bars is made of four bars. Two are abutting. One has a graduated fill of cyan to black; and below it is a black to yellow-brown bar. A thin white filled bar makes the highlight, and all are contained within a black outline. The cigarettes are produced in much the same way – with the addition of the glowing end. The stamped envelopes are three rectangles, one with rounded corners. These are grouped and turned through 25°, then duplicated to create the required number.

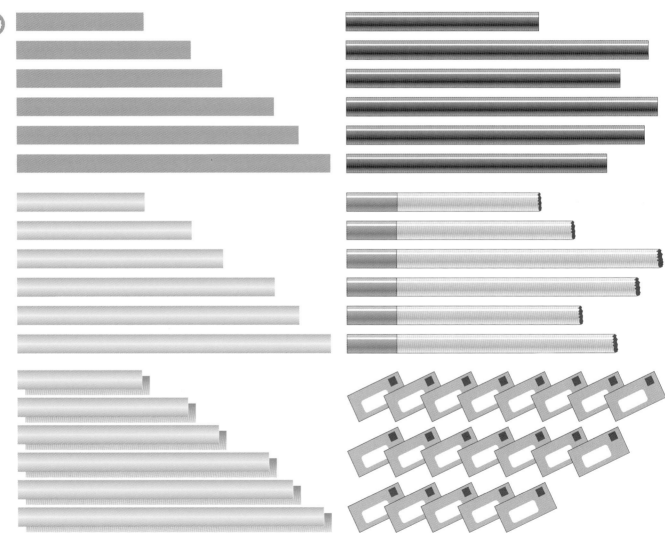

Below These gold
ingots are made up
of four rectangles,
the top face having
a blended fill.
The certificates con-
sist of a blended fill,
plus a circle of red
and some grey lines.
These are grouped
and angled. Vary the
angle to make a more
interesting shape.

Below Spherical
beads are also simple
to create, by using
a circular box with a
circular blended
fill and adding a small
white highlight.
Each of the 3-D bars
is a straightforward
rectangle, with two
skewed rectangles
to form the end and
underside.

You can manufacture
a line of cubes in the
same way, adjusting
the width of the last
cube to create the
fraction.

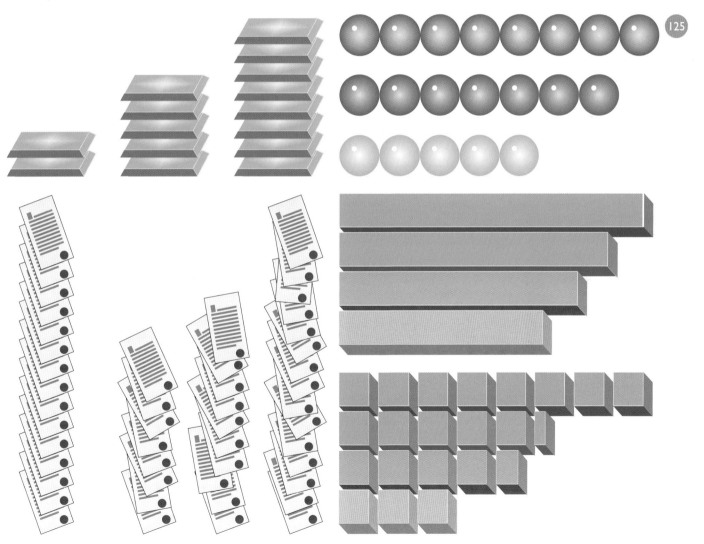

125

PAGE-LAYOUT SOFTWARE 3

Below Square area charts are easy to construct within page-layout programs. Calculate the square root of the value of each item and use this to draw square boxes to the required size. You may then want to resize the boxes as a group to suit your layout.

Below Circular area charts are slightly more complicated. The area of a circle is πr^2 (3.142 x the square of the radius). Divide the value by 3.142 and then find the square root of that figure. This gives you the size of the radius. Double this to obtain the diameter of the circle. Once you've drawn all of the circles you'll probably have to enlarge or reduce them as a group to fit the layout.

Below Unit area charts are even easier to produce. Start with the greatest value and just draw a suitably sized shape, then step and repeat to the required number for each value.

Below Construct a pie chart by drawing a circular box (*1*). Draw a horizontal box from the centre to the edge of the circle. The height of this box should be set to the line width you want (*2*). Take the percentage value of the first segment and multiply it by 3.6 to convert it to degrees. Set the rotation of the new box to that angle (*3*) then move the box so it radiates from the centre of the circle (*4*). Clone the first box again and rotate this by the sum of the first and second segment values (in degrees) (*5*). Next, position this box (*6*) then repeat for all other segments (*7*) except the last (the box for which would be identical to the first). Set the circular box colour and line box colours as you want them (*8*). To add colour to the segments you will need to use a drawing tool to trace over the segment lines. Be sure to extend the segment colour beyond the edge of the circle (*9*). Repeat this, setting the fill colour for each segment (*10*). Arrange the layering of the elements so that the segment-dividing lines are in front (*11*). Finally, duplicate the base circle. Keep it concentric, but extend the diameter so that it encloses all overlapping segment colour boxes. Delete the fill and add a frame of the page colour that is wide enough to clean up the uneven edges (*12*).

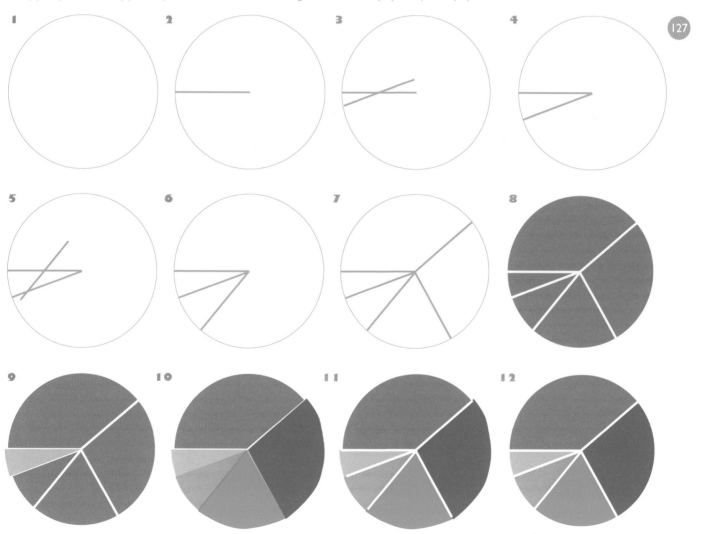

PAGE-LAYOUT SOFTWARE 4

Below You can create the 3-D blocks for organizational diagrams in the same way as 3-D bars. The red dotted lines are simply duplicated circles, as are the drop shadows. Below is a much more subtle treatment, which shows the boxes as embossed plaques.

Below By using a metallic effect, you can give shapes and lines a high-tech look. There are just two boxes for each block, each with a graduated tint. Angled white lines separate the sides to make them look like edges.

Below Convincing spheres, floating above soft shadows, are simple to construct using graduated tints in circular boxes. The shadows are pale shades of grey blended to white.

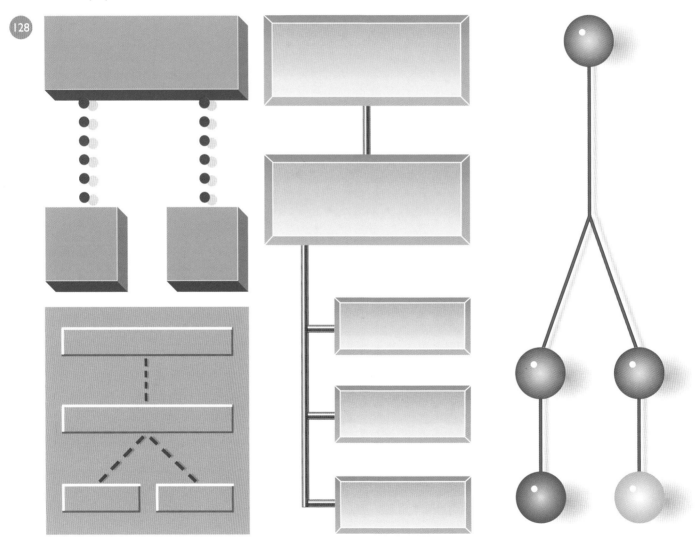

Below You can even create effective maps using a page-layout program. First, import a base map of the required area (a clip-art map would do). Then create a matrix of suitably sized small squares or circles to cover the map area. Delete any shapes that fall outside the map area, and colour those within it as appropriate. The result can be a highly effective graphic. It's also possible, with sufficient memory allocated to software, to create a map from spherical-bead shapes without leaving your page-layout program.

VECTOR SOFTWARE / 1

Undoubtedly, the vector drawing program is the information designer's key tool. Although you can achieve a great deal with a page-layout program and raster painting software, the vector drawing program has been specifically designed to suit the purposes of the graphic designer. A good professional vector program provides minute control combined with a wide range of creative options.

Below left Keep corners of angled boxes tidy by setting the join to either a rounded or a chamfered mitre. Graph lines with acute angles will also benefit from rounding or chamfering.

Below When you're drawing 3-D boxes, the easiest way to get precision at the corners is to clone the first rectangle (that is, duplicate it in the same position), then ungroup the clone and drag the upper corners to the desired position. Simply clone this shape and drag the side corners till they 'snap' to the corners of the original rectangle.

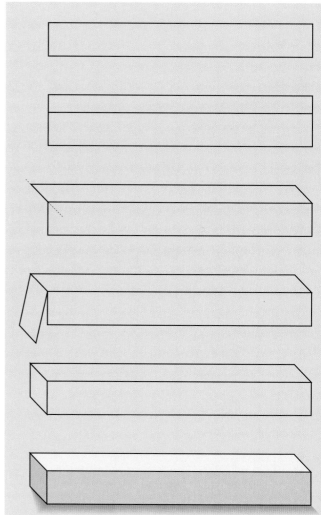

Below Tools like the step-and-repeat facility speed up the diagram-making process. Draw one bar to the proportion you want, then step-and-repeat. Make bar charts by using data values for the box widths, then resize the whole chart to fit the space available.

Below It's possible to create an almost limitless range of line styles with vector drawing software. Multiple graduated fills produce good effects. Try blending cloned lines that are styled differently. Draw wavy lines oversize, then reduce them to produce tight curves. Soft shadows are easily produced by offsetting cloned shapes and filling them with blends from dark to base colour – include the elements of the base colour in the dark shadow specification.

131

VECTOR SOFTWARE / 2

Below Embossed or
engraved effects are
also easy to produce.
The convention is
to assume a virtual
light source at top
left, but don't let that
hinder creativity. So
long as the chart is
readable, you can be
as wild as you like.

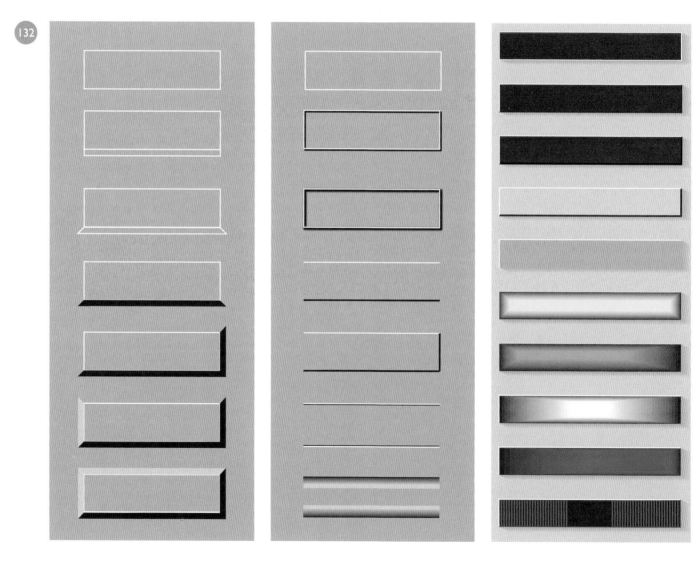

Below Geometric shapes are simple to create and can be given a convincing 3-D quality with a little careful filling and shadowing. Circles with offset graduated fills will look a bit flat. To make spheres, superimpose two circles with blended fills. Follow the steps below. Draw a base circle with no line and a light fill. Add a smaller second circle, offset and with a dark fill. Clone it and blend the original of these with the base circle. Now draw a small circle with a light fill and blend this with the second of the mid circles. This technique works with other shapes, too. You can pick up the top shape and move it to adjust the effect.

Below Use facilities such as joining filled paths to create 'holes' through which a graduated fill or some other complex fill needs to be seen. This is also handy for lake areas in maps.

133

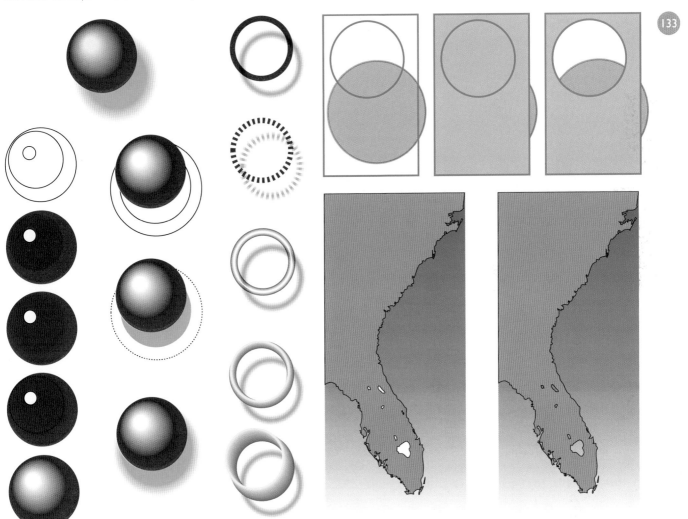

RASTER SOFTWARE

Although they are largely used for modifying or enhancing vector-generated diagrams, it is possible to create charts directly in raster paint programs. You will need to be able to draw shapes and adjust their dimensions, preferably by keying in values rather than by manual manipulation. You will also need to be able to reposition items accurately. But any current version of a raster-based painting program should facilitate that.

You can of course create a base vector image using a page-layout or drawing program and import that into the paint program provided the import has a good level of resolution. The major advantage of using the raster program is in the range of image manipulation features you can use. Photographic images can be used as bases, as they are or with any number of filtered effects, and diagrams can be created from these bases.

Below A colour image serves as the base for this chart. The columns are made by setting the selection tool to the size required. Start by making vertical guides at measured intervals across the image and then set a base line (you'll have to invert the image if the

selection tool selects from the top down). Make the tallest bar first. Note the pixel height and use this as a conversion factor for all the other values in the chart. Reset the selection tool for each column. Once you have all the columns selected you can modify them – by

adding an outline, or adjusting the contrast or the colour choice. If you then select the rest of the image you can manipulate that too.

Below Using the same base a simple graph can be drawn. Use guides to align and draw the grid. The point marker is created and copied. Then each one is located by reference to the grid and a line is drawn to connect the point markers.

Below This column chart was produced simply by cropping a suitable image to the required width and then scaling the part above the runner's head to the required height. Any distortion is immaterial given the blurriness of the background.

Below Pie charts can be constructed from suitable images. Crop the circular shape then draw a line for one side of the first segment. Duplicate the line and rotate it to the degree value required (percentage value × 3.6°). Repeat for the other segment lines. You can then add soft shadow to highlight a segment. Or you can select and move the segments to make an exploded pie chart.

Below Like the bar chart examples on the facing page, this flow chart was made by selecting areas and then adjusting the brightness and contrast settings. Frames were added to the selections and the connecting lines made by filling selected areas.

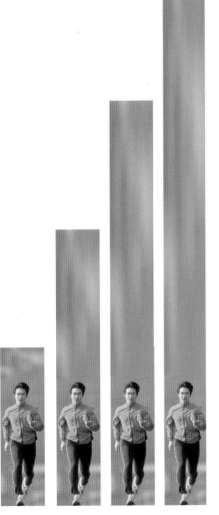

3-D SOFTWARE / 1

You are probably most familiar with those industry-standard applications best suited for producing your day-to-day work – page layout, drawing, image manipulation, web design, and so on. A 3-D application will probably not form part of your graphic armoury, not least because of the steep learning curve required to use one. However, even a basic 3-D application can provide a fast way of creating very effective diagram art, with only a minimal amount of knowledge.

1 | 2 | 3
3-D applications provide tools for generating simple block diagrams in a matter of minutes. This diagram (1, 2) took just ten minutes to construct and render. Size fields allow you to enter accurate dimensions (3).

4 | 5 | 6 | 7 | 8
One immediate advantage of using a 3-D application is that it allows you to quickly rotate an image or alter the perspective to suit particular situations, such as fitting a layout or emphasizing (or disguising!) a statistic.

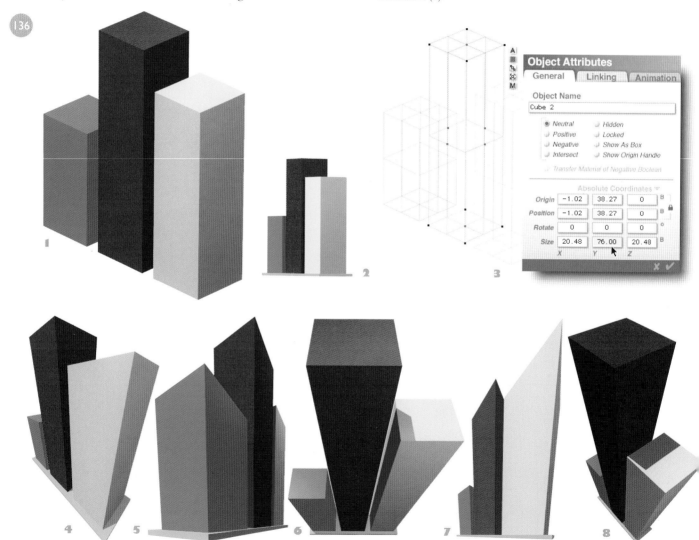

1 | 2 | 3
Most 3-D applications
come with rendering
tools which allow you
to apply a variety of
effects, such as plain
colour (1), solid (2)
or transparent (3)
characteristics which
you can either define
yourself or draw on
from a built-in 'gallery'
of effects.

4 | 5 | 6 | 7
As well as a virtually
limitless variety of
surfaces, objects can
be moulded into any
basic shape, and these
shapes can be
modified, such as this
cylinder (4), which is
shown as a tube (5),
stacked tubes (6) and
tubes within tubes
(7).

137

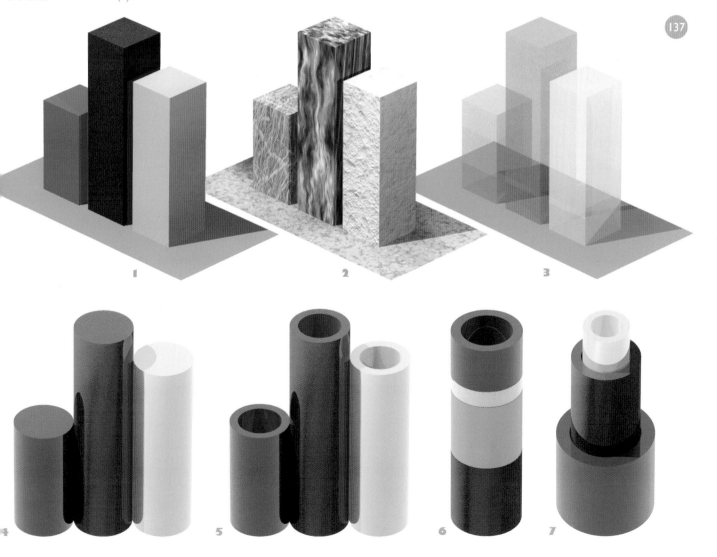

1

2

3

4

5

6

7

3-D SOFTWARE / 2

1 | 2

Although simple techniques can be used to good effect, such as nesting one object within another (*1*) to give comparisons of volume, more dramatic results can be achieved by separating out the objects (*2*).

3

Built-in rendering features give effective graphic results as in example (*2*), but another, often more appropriate technique is to use 3-D 'mapping' features to wrap pictures around objects (*3*).

4 | 5 | 6

The technique of making an indentation in a solid object is known as a 'Boolean' operation, which means, in effect, to subtract one object from another. In the example (*2*), this was achieved by duplicating the cube and then resizing it to the appropriate dimensions (*4*). The smaller cube was then given a 'negative' attribute while the larger one retained its 'positive' one. This operation resulted in removing the smaller cube from the larger one, the inside walls retaining all of the surface and lighting characteristics that had already been set for the larger cube (*5*). Finally, a new cube with a different surface texture was added to the scene, giving it the appearance of hovering proud of the larger cube (*6*, shown end-on).

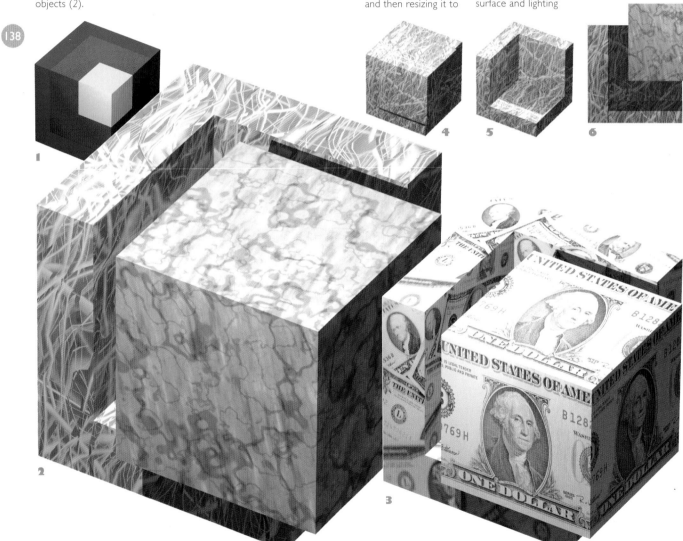

1

2

3

4

5

6

1 | 2

Although 3-D applications provide features for modelling an object out of almost any kind of material, rendering time becomes an important issue for complex objects. The criteria which affect rendering time are effects such as refraction, reflection and transparency – the more you apply of each the longer a scene takes to render. The glass discs in the pie diagram (*1*) use high settings of all three attributes and consequently took around twelve hours to render at this size – even on a fast computer! Diagram (*2*) uses the same discs as (*1*) but with simpler textures applied and thus rendered in a matter of minutes rather than hours.

3

Another advantage of 3-D rendering applications is that they provide an effective way of departing from conventional methods of representing values, as shown here with these glass balls of varying sizes.

2

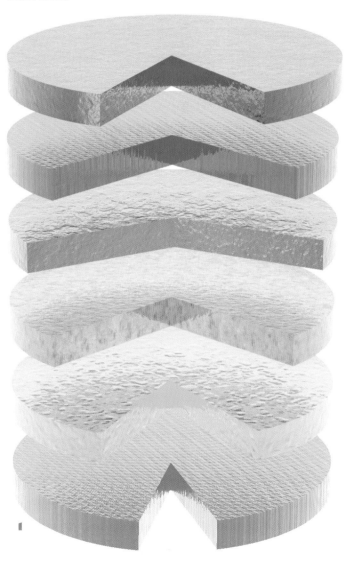

1

3

READY-MADE RESOURCES

Ready-made graphic resources in the form of clip-art collections, royalty-free photographs and backgrounds abound, with a quality of image ranging from high-end professional (at high-end cost) to unusable – and all too often the latter. If you are a professional graphic designer you may scorn the idea of such ready-made resources on the basis that their accessibility runs contrary to the ethos of your profession – after all, you're being paid to provide originality, not merely to

regurgitate the same material that anyone can use. However, image resources need not be viewed as competition for your imagination, but as an extension to it, particularly if it provides a cost-effective means of incorporating otherwise costly images into your graphics.

Perhaps the most valuable resources are high-end mapping resources such as Cartesia's *Map Art* and Digital Wisdom's *Mountain High Maps*, the latter of which is used to illustrate these pages. Generating

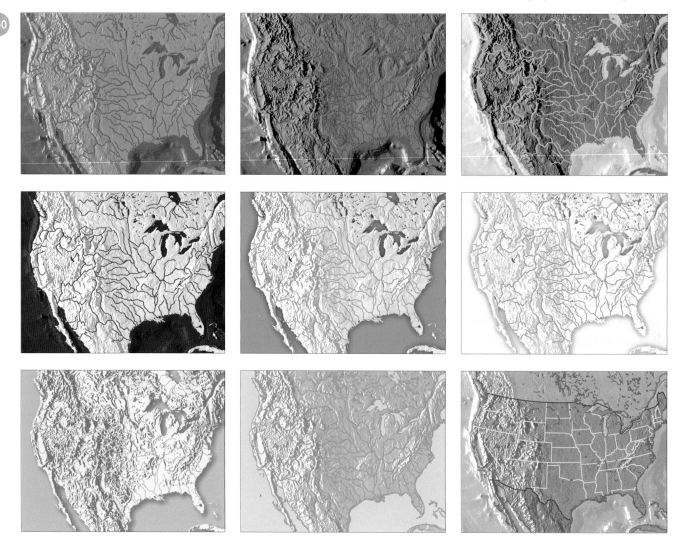

accurate maps can be a costly and time-consuming business, often getting in the way of your creative flow, not to mention deadlines. These resources generally come in the form of a 'kit' of separate parts (topographic relief, coastlines, rivers, country borders and so on) which not only allows you to select just the elements that you need for the job, but to use the drawing or image-manipulation applications with which you are already familiar. The rest is up to your imagination.

Below Mapping resources generally come as a 'kit' of separate parts, provided as separate layers or individual components in industry-standard file formats, thus allowing you to use the software packages with which you are most familiar. The examples below were generated in Adobe Photoshop, using 'alpha channels' for land and sea masks, and 'paths' for rivers and state borders. The wide variety of effects were applied simply by playing with image adjustment controls and shadow effects.

141

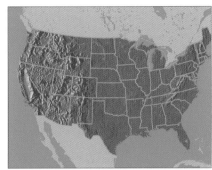

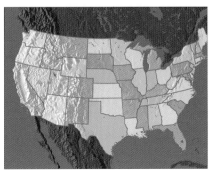

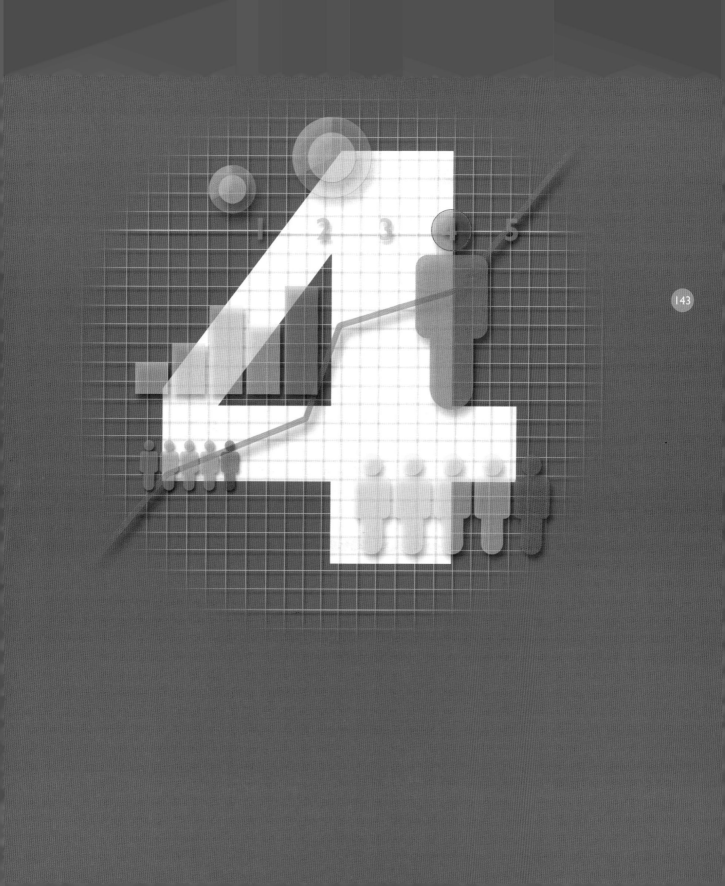

Grundy & Northedge
Heading icons for
Bloomberg magazine.
Created with Adobe
Illustrator. Unusually
detailed icons covering
a wide range of
complex activities.
The underlying grid
structure and consistent
graphic treatment
unifies what might
otherwise be a
disparate collection
of shapes and styles.

144

Strategies

Web watch

Tuned in

Toolbox

Signals

What's working

Books

That was then

Getting technical

Grundy & Northedge
Icons for the Camelot
annual report (designed
by MPL). Created using
Adobe Illustrator.
In this series colour is
used to hold icon
groupings in their
relevant categories.

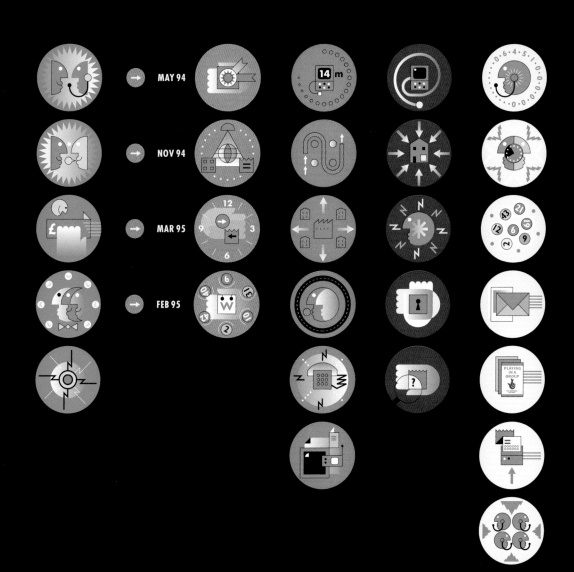

Nigel Holmes
Illustration showing
readers how to
carry out a breast
examination, created
for *Self* magazine
using Macromedia
FreeHand. The human
figure is reduced to
a simple, stylized
outline.

146

Annabel Milne
How to clean your teeth. Created for demonstration purposes using Adobe Illustrator. This realistic styling of the human form still succeeds in keeping detail to a minimum. The sequence shows line and tone methods varying from monochrome with line and flat tint through to simple colour rendering.

147

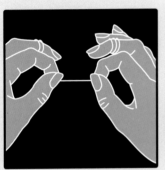
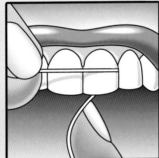
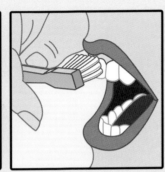
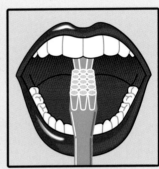

Nigel Holmes
Illustration of Greg Louganis' reverse two-and-a-half-pike dive, created for *Olympic Dreams* (published by Rizzoli) using Macromedia FreeHand. A complex physical action clearly shown in a simple sillhouette sequence.

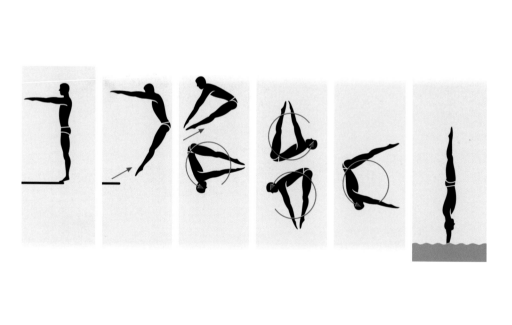

Annabel Milne
Angiogenesis, showing payloaded antibodies targeting cancer cells. Created for Antisoma plc prospectus, using Adobe Illustrator and Adobe Dimensions. The exaggerated perspective view and bright colours on a black base gives the diagram a dramatic space-fantasy feel. The various items cannot, of course, be drawn realistically but instead have an almost mechanical styling, which also contributes to the overall effect.

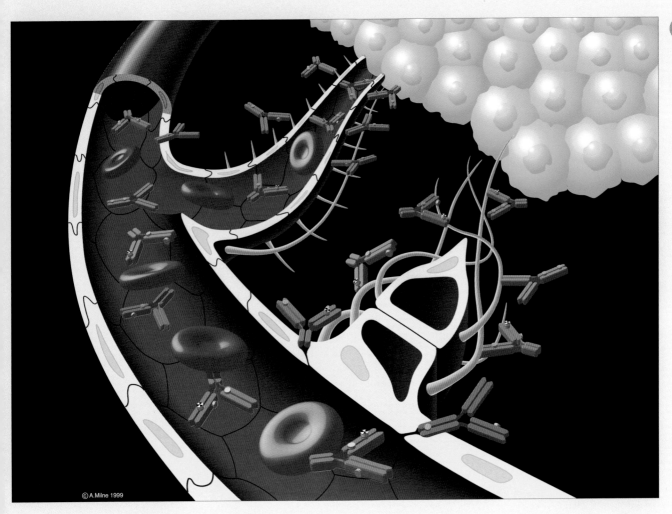

© A.Milne 1999

Nigel Holmes
Map of the parts of the brain, created for a medical magazine published by *Time*, using Macromedia FreeHand. This highly stylized modular treatment uses coloured type to identify and describe the component parts.

Although created with a vector drawing program, a treatment like this could be achieved with page layout software.

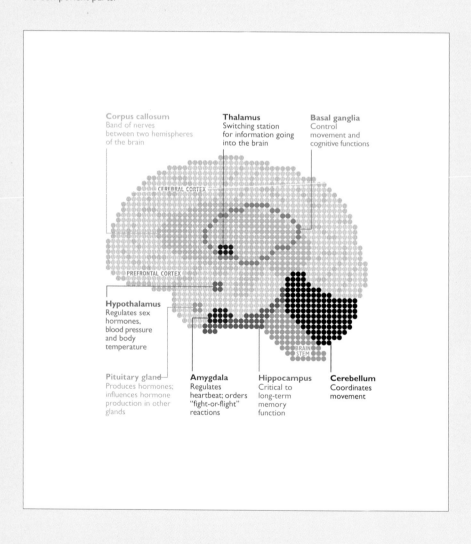

Corpus callosum
Band of nerves between two hemispheres of the brain

Thalamus
Switching station for information going into the brain

Basal ganglia
Control movement and cognitive functions

CEREBRAL CORTEX

PREFRONTAL CORTEX

Hypothalamus
Regulates sex hormones, blood pressure and body temperature

BRAIN STEM

Pituitary gland
Produces hormones; influences hormone production in other glands

Amygdala
Regulates heartbeat; orders "fight-or-flight" reactions

Hippocampus
Critical to long-term memory function

Cerebellum
Coordinates movement

John Grimwade
Round-the-world balloons, created for *Popular Science* using Adobe Illustrator. Two different scales are employed in this illustrative diagram. One is used to show the comparative sizes of the balloons, the other for the cabins. Each has a scale bar but the second also has the human sillhouettes that help in establishing a visual scale.

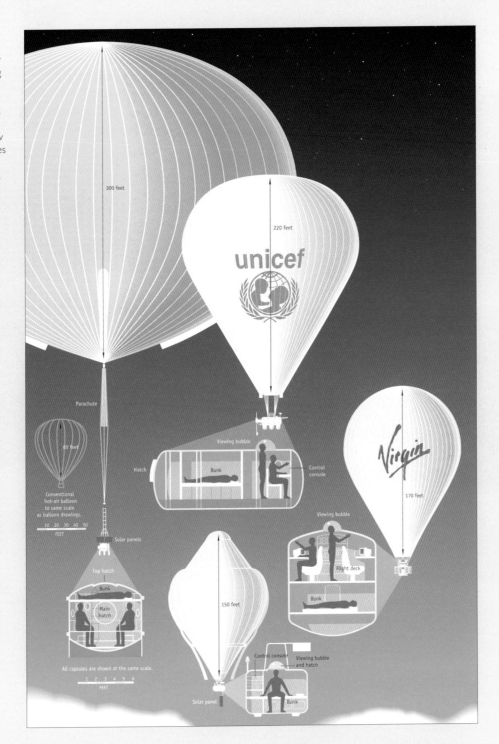

Nigel Holmes
Illustration of the
workings of a gas
engine, created for
Attaché magazine
using Macromedia
FreeHand. A clever
combination of
detailed technical
representation with
a cartoon treatment
to lighten the effect.

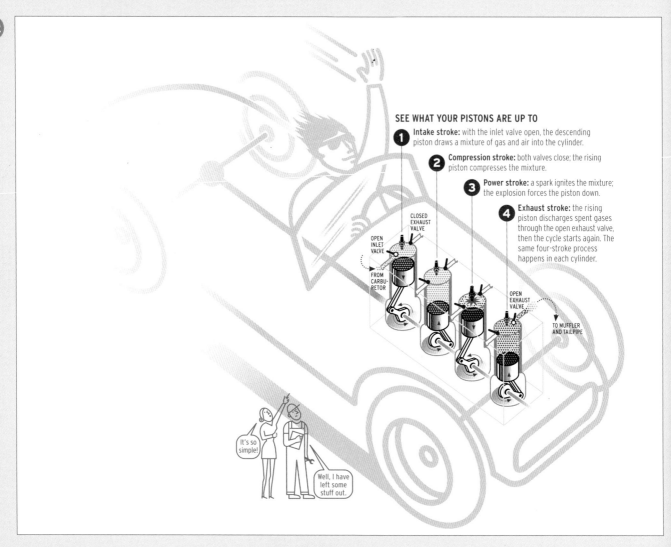

Bill Le Bihan
The 4-stroke combustion engine cycle (known affectionately as suck, squeeze, bang, puff), created for a partwork series using Macromedia FreeHand. A more detailed diagram of the combustion process making good use of the blend feature available in vector drawing software.

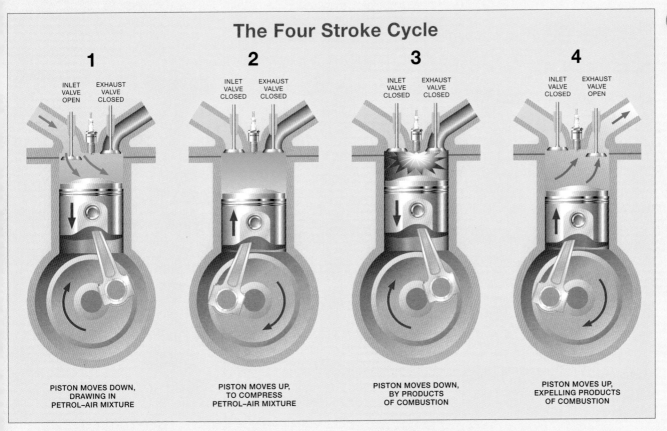

The Four Stroke Cycle

1

INLET
VALVE
OPEN

EXHAUST
VALVE
CLOSED

PISTON MOVES DOWN,
DRAWING IN
PETROL–AIR MIXTURE

2

INLET
VALVE
CLOSED

EXHAUST
VALVE
CLOSED

PISTON MOVES UP,
TO COMPRESS
PETROL–AIR MIXTURE

3

INLET
VALVE
CLOSED

EXHAUST
VALVE
CLOSED

PISTON MOVES DOWN,
BY PRODUCTS
OF COMBUSTION

4

INLET
VALVE
CLOSED

EXHAUST
VALVE
OPEN

PISTON MOVES UP,
EXPELLING PRODUCTS
OF COMBUSTION

153

John Grimwade
Illustration showing how a pipe organ works, created for the *Smithsonian* magazine using Adobe Illustrator. Several diagrams are successfully combined here to illustrate a highly complex structure and process. In order to achieve this many of the pipes are not shown and other elements are reduced to simple structures. Note the use of the human silhouette to aid appreciation of the scale of the pipes.

154

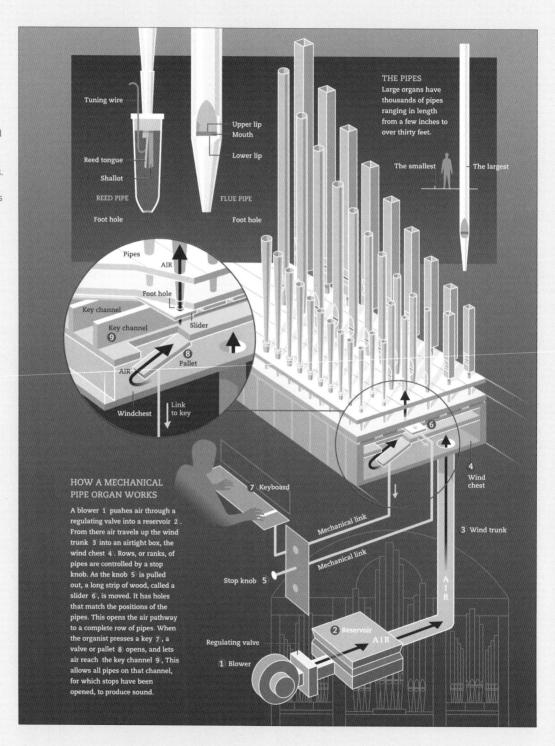

Tuning wire

Reed tongue

Shallot

REED PIPE

Foot hole

Upper lip
Mouth

Lower lip

FLUE PIPE

Foot hole

THE PIPES
Large organs have thousands of pipes ranging in length from a few inches to over thirty feet.

The smallest   The largest

Pipes

AIR

Foot hole

Key channel

Key channel
9

AIR

Slider

8
Pallet

Windchest

Link
to key

HOW A MECHANICAL
PIPE ORGAN WORKS

A blower 1 pushes air through a regulating valve into a reservoir 2 . From there air travels up the wind trunk 3 into an airtight box, the wind chest 4 . Rows, or ranks, of pipes are controlled by a stop knob. As the knob 5 is pulled out, a long strip of wood, called a slider 6 , is moved. It has holes that match the positions of the pipes. This opens the air pathway to a complete row of pipes. When the organist presses a key 7 , a valve or pallet 8 opens, and lets air reach the key channel 9 . This allows all pipes on that channel, for which stops have been opened, to produce sound.

7 Keyboard

6

4
Wind
chest

Mechanical link

3 Wind trunk

Mechanical link

Stop knob 5

AIR

Regulating valve

2 Reservoir

AIR

1 Blower

John Grimwade
Cutaway diagram of New York's Grand Central Station, created for *Condé Nast Traveler* using Adobe Illustrator. Another highly complex structure edited to facilitate understanding by the lay-reader. Floors are 'floated' apart to reveal information. The human forms are used here again to indicate the scale of the structure.

155

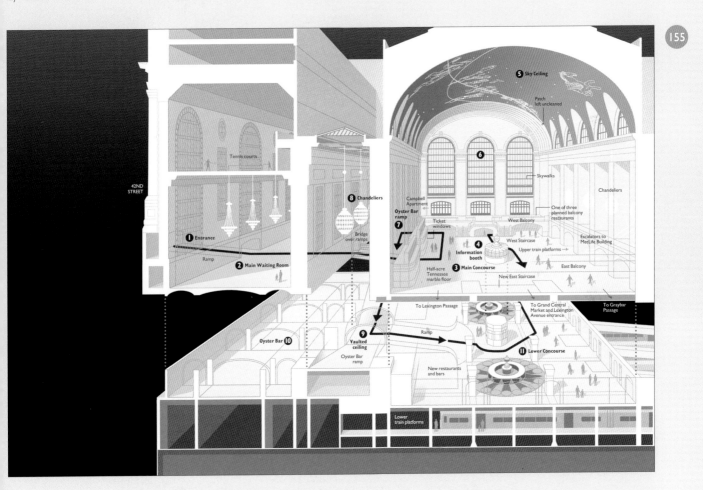

42ND STREET

Tennis courts

Chandeliers ⑧

⑤ Sky Ceiling

Patch left uncleaned

Skywalks

Chandeliers

Campbell Apartment

Oyster Bar ramp ⑦

Ticket windows

West Balcony

One of three planned balcony restaurants

⑥

Bridge over ramps

West Staircase

Escalators to MetLife Building

① Entrance

Information booth ④

Upper train platforms →

Ramp

② Main Waiting Room

Half-acre Tennessee marble floor

③ Main Concourse

East Balcony

New East Staircase

To Lexington Passage

To Grand Central Market and Lexington Avenue entrance

To Graybar Passage

Oyster Bar ⑩

⑨ Vaulted ceiling

Oyster Bar ramp

Ramp

New restaurants and bars

⑪ Lower Concourse

Lower train platforms

Nigel Holmes
Illustration explaining the size of the US national debt. Created for Richard Saul Wurman's book *Understanding*, using Macromedia FreeHand. A series of charts and illustrations that serve as a visual analogy ('a drop in the ocean'). Despite the quirky treatment, the hard facts are illustrated with great clarity.

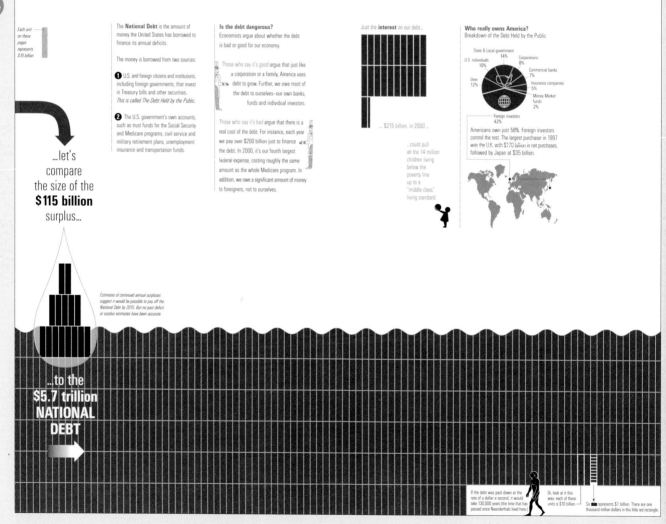

bounford.com
Statistical bar
chart created for
Worldlink magazine,
using Macromedia
FreeHand. The grid is
used here as a visual
benchmark with
excessive items
breaking out beyond
the general norm.
Even so there are
values too great to be
included in the scale
and these have to be
represented by the
'broken' bar.

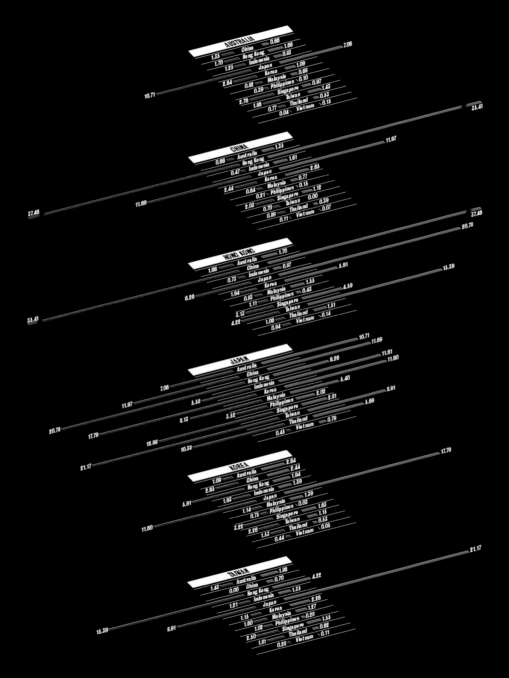

Grundy & Northedge
Five financial crashes in history, created for *Worldlink* magazine using Adobe Illustrator. A light treatment using the visual analogy of bursting bubbles – the pin indicating the point in time when the bubbles burst.

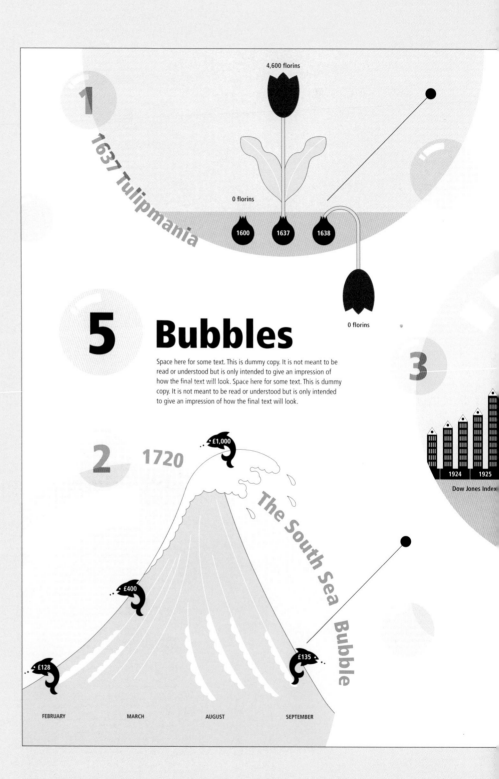

159

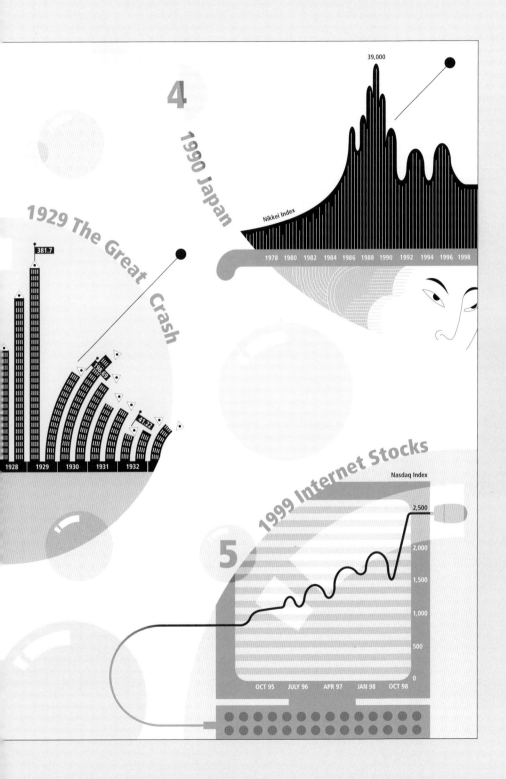

4

1990 Japan

39,000

Nikkei Index

1978 1980 1982 1984 1986 1988 1990 1992 1994 1996 1998

1929 The Great Crash

381.7

188.57

41.22

1928 1929 1930 1931 1932

1999 Internet Stocks

5

Nasdaq Index

2,500

2,000

1,500

1,000

500

0

OCT 95 JULY 96 APR 97 JAN 98 OCT 98

bounford.com
Diagram showing
the range of
airport investment
by BAA. Created
using Macromedia
FreeHand. Using the
map base to locate
the airports helps
the reader see the
regional allocation
of investment.

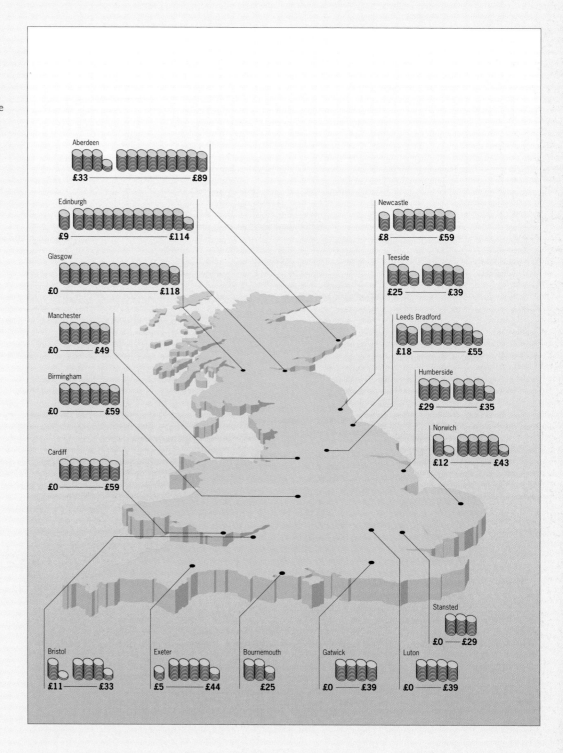

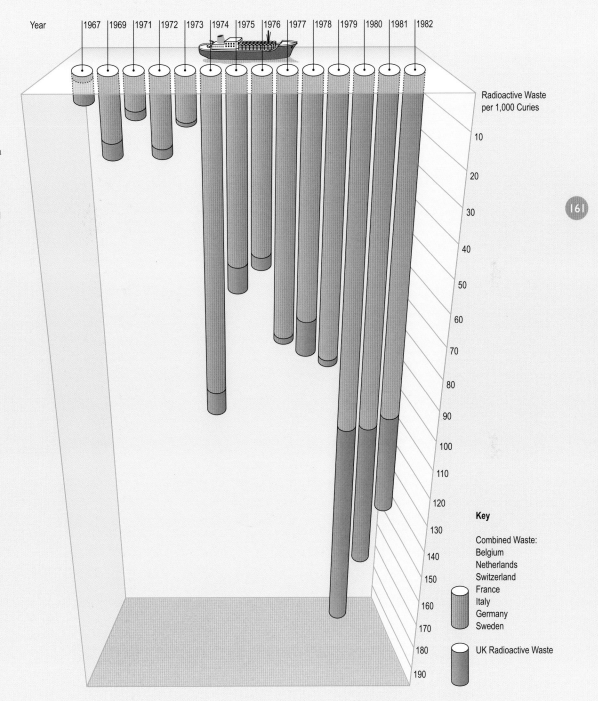

bounford.com
Diagram showing the dumping of radioactive waste at sea by European countries, created for Greenpeace International using Macromedia FreeHand. A simple bar chart given drama by the illustrative 'background'. The use of perspective enhances the effect but a scale is included to enable the values to be read accurately.

Year

1967 1969 1971 1972 1973 1974 1975 1976 1977 1978 1979 1980 1981 1982

Radioactive Waste per 1,000 Curies

10
20
30
40
50
60
70
80
90
100
110
120
130
140
150
160
170
180
190

161

Key

Combined Waste:
Belgium
Netherlands
Switzerland
France
Italy
Germany
Sweden

UK Radioactive Waste

John Grimwade
Map showing the layout and location of Angkor Wat, created for *Condé Nast Traveler* using Adobe Illustrator. A simple street map drawn in perspective which is enhanced by the atmospheric aerial perspective effect.

Salient buildings are shown as 3-D illustrations. The inset location map is drawn to the same perspective projection.

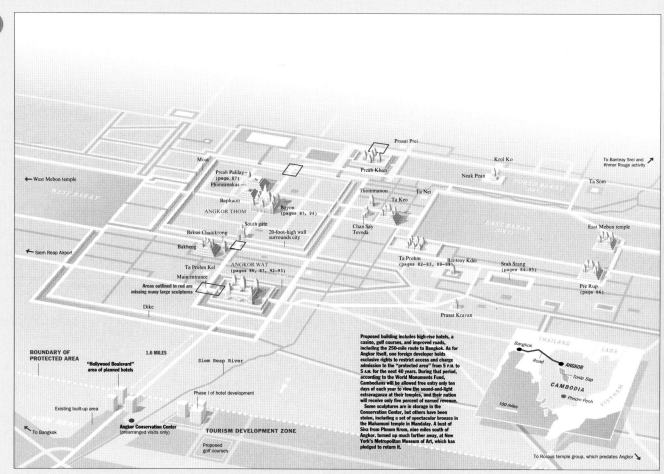

Prasat Prei

Krol Ko

To Banteay Srei and Khmer Rouge activity

Moat

Preah Khan

Neak Pean

Ta Som

← West Mebon temple

Preah Palilay— (page 87) Phimeanakas

Thommanon

Ta Nei

WEST BARAY

Baphuon

ANGKOR THOM

Bayon (pages 87, 94)

Ta Keo

Chan Say Tevoda

East Mebon temple

Baksei Chamkrong

South gate

20-foot-high wall surrounds city

← Siem Reap Airport

Bakheng

Ta Prohm (pages 82–83, 88–89)

Banteay Kdei

Srah Srang (pages 84–85)

Ta Prohm Kel

ANGKOR WAT (pages 86, 87, 92–93)

Main entrance

Areas outlined in red are missing many large sculptures

Pre Rup (page 86)

Dike

Prasat Kravan

BOUNDARY OF PROTECTED AREA

1.6 MILES

Siem Reap River

Proposed building includes high-rise hotels, a casino, golf courses, and improved roads, including the 250-mile route to Bangkok. As for Angkor itself, one foreign developer holds exclusive rights to restrict access and charge admission to the "protected area" from 5 P.M. to 5 A.M. for the next 40 years. During that period, according to the World Monuments Fund, Cambodians will be allowed free entry only ten days of each year to view the sound-and-light extravaganza at their temples, and their nation will receive only five percent of earned revenue.
Some sculptures are in storage in the Conservation Center, but others have been stolen, including a set of spectacular bronzes in the Mahamuni temple in Mandalay. A bust of Siva from Phnom Krom, nine miles south of Angkor, turned up much farther away, at New York's Metropolitan Museum of Art, which has pledged to return it.

THAILAND

LAOS

Bangkok

Road

ANGKOR

"Hollywood Boulevard" area of planned hotels

Phase I of hotel development

Tonle Sap

CAMBODIA

VIETNAM

Existing built-up area

Phnom Penh

100 miles

← To Bangkok

Angkor Conservation Center (prearranged visits only)

TOURISM DEVELOPMENT ZONE

Proposed golf courses

To Rolous temple group, which predates Angkor →

Nigel Holmes
The bombing of the
Uffizzi Museum in
Florence, created for
Time magazine
using Macromedia
FreeHand. In this
news diagram, a
relatively simple 3-D
representation of the
building is shown
superimposed on

the local street layout.
Labels are kept away
from the key
structure and are
located by arrows.

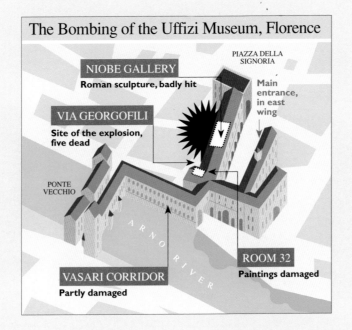

The Bombing of the Uffizi Museum, Florence

NIOBE GALLERY
Roman sculpture, badly hit

PIAZZA DELLA SIGNORIA

Main entrance, in east wing

VIA GEORGOFILI
Site of the explosion, five dead

PONTE VECCHIO

ARNO RIVER

VASARI CORRIDOR
Partly damaged

ROOM 32
Paintings damaged

John Grimwade
Map of central Florence, created for *Condé Nast Traveler* using Adobe Illustrator. A basic street plan given a very effective 3-D treatment by the addition of simplified representations of key buildings, enhanced by cast shadows. Shadows are also used to great effect under the bridges.

164

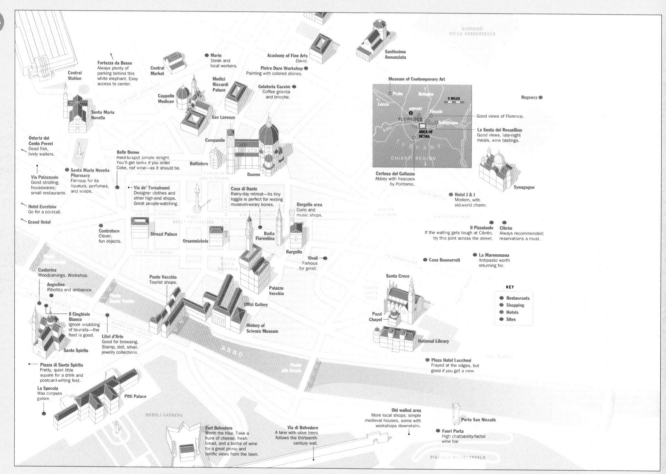

Grundy & Northedge
A movie-goer's view of San Francisco, created for the *Sunday Times* using Adobe Illustrator. A highly stylized treatment of a street map – although the grid layout of many American city streets can hardly be treated otherwise. This map is enhanced by the addition of silhouettes of key structures. The near geometric representation of the coastline gives an informal feel to the map.

bounford.com
Flow chart showing
vehicle loading
and offloading at
the Channel Tunnel
terminals, created
for Eurotunnel plc
using Macromedia
FreeHand. A
simplified map base
serves to orientate
travellers embarking

and disembarking
at either end of the
tunnel. The function
of the diagram is to
reassure passengers
by demonstrating
the simplicity of the
procedures.

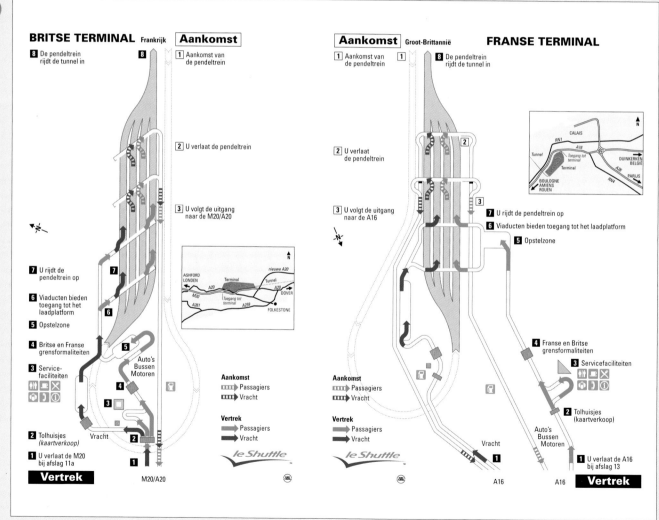

John Grimwade
Diagram describing how hurricanes happen, created for *Popular Science* using Adobe Illustrator. A fascinating composite of several diagrams – relational, statistical and illustrative. This page demonstrates how well text and imagery can be integrated to make an engrossing information tableau.

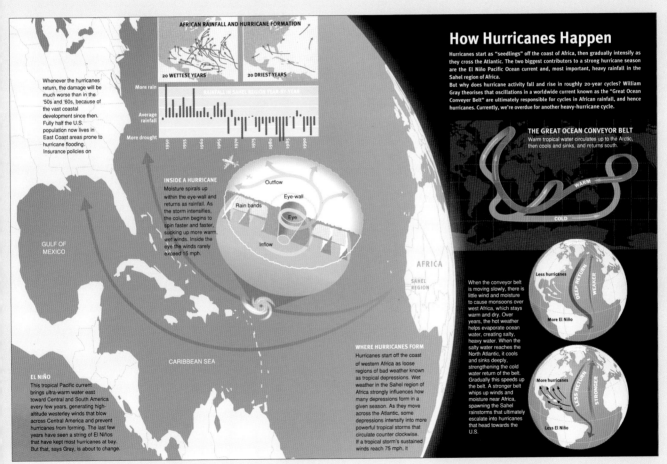

167

AFRICAN RAINFALL AND HURRICANE FORMATION

20 WETTEST YEARS

20 DRIEST YEARS

More rain

Average rainfall

More drought

RAINFALL IN SAHEL REGION YEAR-BY-YEAR

1950 1955 1960 1965 1970 1975 1980 1985 1990

Whenever the hurricanes return, the damage will be much worse than in the '50s and '60s, because of the vast coastal development since then. Fully half the U.S. population now lives in East Coast areas prone to hurricane flooding. Insurance policies on

INSIDE A HURRICANE
Moisture spirals up within the eye-wall and returns as rainfall. As the storm intensifies, the column begins to spin faster and faster, sucking up more warm, wet winds. Inside the eye the winds rarely exceed 15 mph.

Outflow

Eye-wall

Rain bands

Eye

Inflow

GULF OF MEXICO

EL NIÑO
This tropical Pacific current brings ultra-warm water east toward Central and South America every few years, generating high-altitude westerly winds that blow across Central America and prevent hurricanes from forming. The last few years have seen a string of El Niños that have kept most hurricanes at bay. But that, says Gray, is about to change.

CARIBBEAN SEA

AFRICA

SAHEL REGION

WHERE HURRICANES FORM
Hurricanes start off the coast of western Africa as loose regions of bad weather known as tropical depressions. Wet weather in the Sahel region of Africa strongly influences how many depressions form in a given season. As they move across the Atlantic, some depressions intensify into more powerful tropical storms that circulate counter clockwise. If a tropical storm's sustained winds reach 75 mph, it

How Hurricanes Happen

Hurricanes start as "seedlings" off the coast of Africa, then gradually intensify as they cross the Atlantic. The two biggest contributors to a strong hurricane season are the El Niño Pacific Ocean current and, most important, heavy rainfall in the Sahel region of Africa.

But why does hurricane activity fall and rise in roughly 20-year cycles? William Gray theorises that oscillations in a worldwide current known as the "Great Ocean Conveyer Belt" are ultimately responsible for cycles in African rainfall, and hence hurricanes. Currently, we're overdue for another heavy-hurricane cycle.

THE GREAT OCEAN CONVEYOR BELT
Warm tropical water circulates up to the Arctic, then cools and sinks, and returns south.

WARM

COLD

When the conveyor belt is moving slowly, there is little wind and moisture to cause monsoons over west Africa, which stays warm and dry. Over years, the hot weather helps evaporate ocean water, creating salty, heavy water. When the salty water reaches the North Atlantic, it cools and sinks deeply, strengthening the cold water return of the belt. Gradually this speeds up the belt. A stronger belt whips up winds and moisture near Africa, spawning the Sahel rainstorms that ultimately escalate into hurricanes that head towards the U.S.

Less hurricanes

DEEP RETURN

WEAKER

More El Niño

More hurricanes

LESS RETURN

STRONGER

Less El Niño

Rowson Holbrook
Map showing railway-track modifications on the North London line. Designed for the European Passenger Service. Created using Macromedia FreeHand. The point of the diagram was to show the intricate complexity of re-engineering works on a busy section of London's overground railway system.

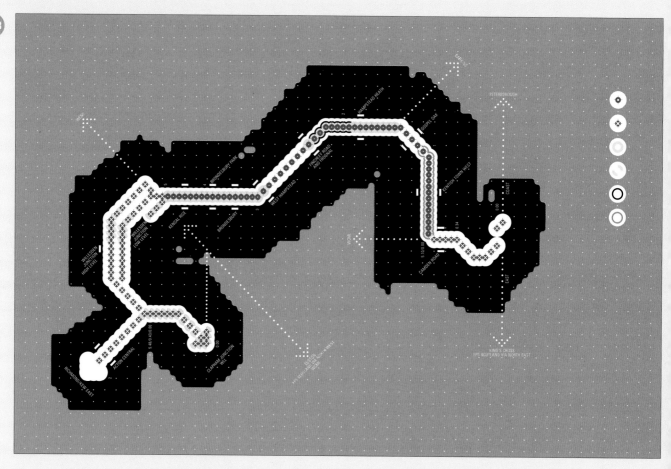

John Grimwade
Diagram and maps describing a damming project on China's Yangtze River, created for *Condé Nast Traveler* using Adobe Illustrator. Four map styles are employed here to give the reader a clear locational context and to indicate the scale of the project. In the illustrative diagram a common element – the boat – is used to convey the enormity of scale.

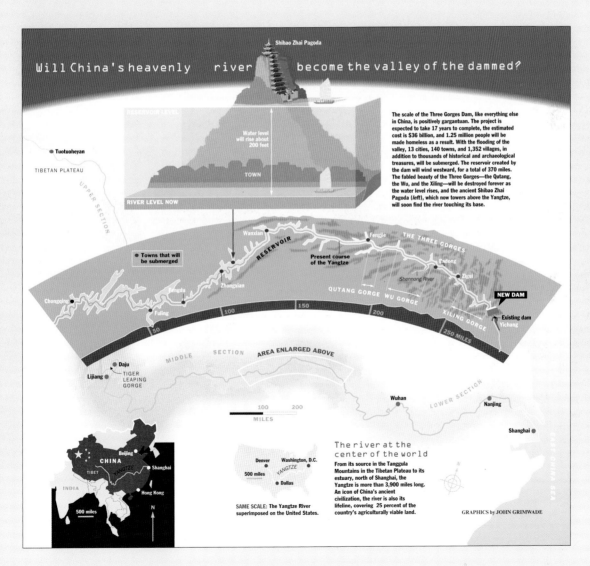

Will China's heavenly river become the valley of the dammed?

Shibao Zhai Pagoda

RESERVOIR LEVEL

Water level will rise about 200 feet

TOWN

RIVER LEVEL NOW

The scale of the Three Gorges Dam, like everything else in China, is positively gargantuan. The project is expected to take 17 years to complete, the estimated cost is $36 billion, and 1.25 million people will be made homeless as a result. With the flooding of the valley, 13 cities, 140 towns, and 1,352 villages, in addition to thousands of historical and archaeological treasures, will be submerged. The reservoir created by the dam will wind westward, for a total of 370 miles. The fabled beauty of the Three Gorges—the Qutang, the Wu, and the Xiling—will be destroyed forever as the water level rises, and the ancient Shibao Zhai Pagoda (*left*), which now towers above the Yangtze, will soon find the river touching its base.

Tuotuoheyan

TIBETAN PLATEAU

UPPER SECTION

● Towns that will be submerged

RESERVOIR

Present course of the Yangtze

THE THREE GORGES

Wanxian

Fengjie

Badong

Zigui

Shennong River

NEW DAM

QUTANG GORGE   WU GORGE

Chongqing

Fengdu

Zhongxian

Fuling

50

100

150

200

XILING GORGE

250 MILES

Existing dam Yichang

MIDDLE   SECTION   AREA ENLARGED ABOVE

Daju
Lijiang
TIGER LEAPING GORGE

Wuhan

LOWER SECTION

Nanjing

100   200
MILES

Shanghai

The river at the center of the world

Denver   Washington, D.C.

500 miles   YANGTZE

Dallas

From its source in the Tanggula Mountains in the Tibetan Plateau to its estuary, north of Shanghai, the Yangtze is more than 3,900 miles long. An icon of China's ancient civilization, the river is also its lifeline, covering 25 percent of the country's agriculturally viable land.

SAME SCALE: The Yangtze River superimposed on the United States.

GRAPHICS by JOHN GRIMWADE

CHINA
Beijing
Shanghai
TIBET   YANGTZE
INDIA   Hong Kong
500 miles   N

EAST CHINA SEA

Nigel Holmes
A double-page spread from Richard Saul Wurman's book *Understanding*, illustrating US Federal Budget income and expenditure. The book is composed entirely of graphic spreads and contains no text. Created using Macromedia FreeHand. Skilful use of icons aid the reader in assimilating a vast amount of detailed statistical information.

170

Individual income tax
$899.7 billion

The tax levied on your salary and any other income you have, such as profits made on investments or interest earned on savings. It's a "progressive" tax, which means that the higher your income the greater percentage of it you'll pay to the government. Tax rates range from 15% to 39.6%.

47.8%

Corporate income tax
$189.4 billion

A tax on corporate income. Rates range from 15% to 38%.

10.1%

Social insurance tax (FICA)
$636.5 billion

The Social Security and Medicare tax, which is withheld from your salary. The rate is 15%; half is paid by you and half by your employer. The self-employed pay the full 15% of their income themselves.

33.8%

Excise tax
$69.9 billion

A tax on goods such as tobacco and alcohol.

3.7%

Estate and gift tax
$27.0 billion

Estate tax is levied on property at the time of death. Gift tax is due on large gifts.

Customs duty
$18.4 billion

A tax on certain imports coming into the U.S.

Miscellaneous receipts
$42.1 billion

The Federal government's total income for 2000 is projected to be **$1.9 trillion**

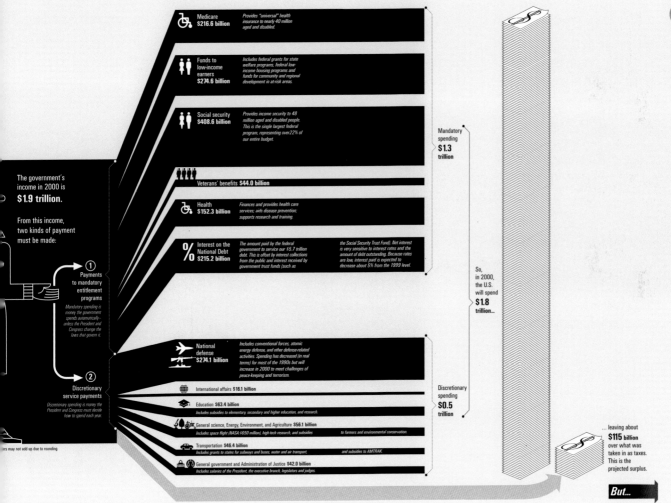

The government's income in 2000 is **$1.9 trillion.**

From this income, two kinds of payment must be made:

① **Payments to mandatory entitlement programs**

Mandatory spending is money the government spends automatically--unless the President and Congress change the laws that govern it.

② **Discretionary service payments**

Discretionary spending is money the President and Congress must decide how to spend each year.

rs may not add up due to rounding

Medicare $216.6 billion
Provides "universal" health insurance to nearly 40 million aged and disabled.

Funds to low-income earners $274.6 billion
Includes federal grants for state welfare programs, federal low-income housing programs and funds for community and regional development in at-risk areas.

Social security $408.6 billion
Provides income security to 48 million aged and disabled people. This is the single largest federal program, representing over 22% of our entire budget.

Veterans' benefits $44.0 billion

Health $152.3 billion
Finances and provides health care services; aids disease prevention; supports research and training.

Interest on the National Debt $215.2 billion
The amount paid by the federal government to service our $5.7 trillion debt. This is offset by interest collections from the public and interest received by government trust funds (such as the Social Security Trust Fund). Net interest is very sensitive to interest rates and the amount of debt outstanding. Because rates are low, interest paid is expected to decrease about 5% from the 1999 level.

National defense $274.1 billion
Includes conventional forces, atomic energy defense, and other defense-related activities. Spending has decreased (in real terms) for most of the 1990s but will increase in 2000 to meet challenges of peace-keeping and terrorism.

International affairs $16.1 billion

Education $63.4 billion
Includes subsidies to elementary, secondary and higher education, and research.

General science, Energy, Environment, and Agriculture $56.1 billion
Includes space flight (NASA $650 million), high-tech research, and subsidies to farmers and environmental conservation.

Transportation $46.4 billion
Includes grants to states for subways and buses, water and air transport, and subsidies to AMTRAK.

General government and Administration of Justice $42.0 billion
Includes salaries of the President, the executive branch, legislators and judges.

Mandatory spending **$1.3 trillion**

Discretionary spending **$0.5 trillion**

So, in 2000, the U.S. will spend **$1.8 trillion...**

... leaving about **$115 billion** over what was taken in as taxes. This is the projected surplus.

But...

bounford.com
Chart showing the global organization of advertising agencies. Created for *The Economist* using Macromedia FreeHand. Colour is used to identify 'family' groupings with colours blended to indicate joint interests. A chart like this will take many hours of preparation in order to resolve successfully all the inter-relationships.

172

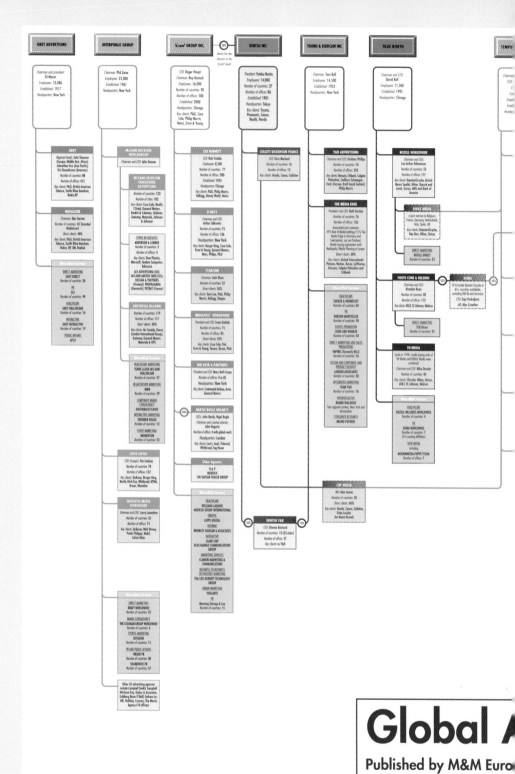

Global A

Published by M&M Euro

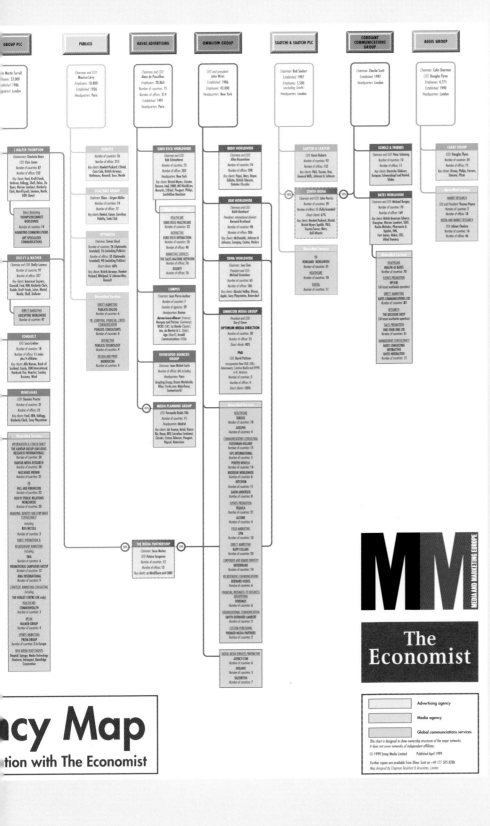

John Grimwade
Diagrammatic map describing the North Atlantic air traffic system, created for *Condé Nast Traveler* using Adobe Illustrator. A highly complex combination of relational and organizational diagrams. The use of perspective and shading lends the chart realism.

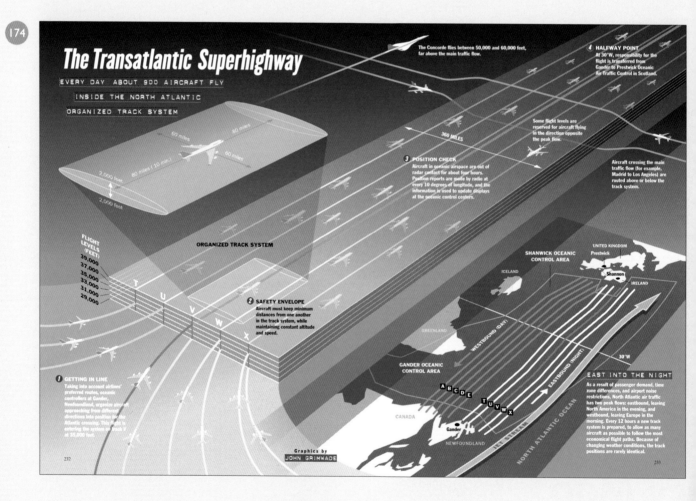

The Transatlantic Superhighway

EVERY DAY ABOUT 900 AIRCRAFT FLY

INSIDE THE NORTH ATLANTIC

ORGANIZED TRACK SYSTEM

The Concorde flies between 50,000 and 60,000 feet, far above the main traffic flow.

HALFWAY POINT
At 30°W, responsibility for the flight is transferred from Gander to Prestwick Oceanic Air Traffic Control in Scotland.

Some flight levels are reserved for aircraft flying in the direction opposite the peak flow.

360 MILES

POSITION CHECK
Aircraft in oceanic airspace are out of radar contact for about four hours. Position reports are made by radio at every 10 degrees of longitude, and the information is used to update displays at the oceanic control centers.

Aircraft crossing the main traffic flow (for example, Madrid to Los Angeles) are routed above or below the track system.

60 miles

80 miles

60 miles

60 miles (10 min.)

2,000 feet

2,000 feet

ORGANIZED TRACK SYSTEM

FLIGHT LEVELS (FEET)
39,000
37,000
35,000
33,000
31,000
29,000

T
U
V
W
X

SAFETY ENVELOPE
Aircraft must keep minimum distances from one another in the track system, while maintaining constant altitude and speed.

UNITED KINGDOM
Prestwick

SHANWICK OCEANIC CONTROL AREA

ICELAND

Shannon

IRELAND

GREENLAND

WESTBOUND (DAY)

EASTBOUND (NIGHT)

30°W

GANDER OCEANIC CONTROL AREA

A B C D E
T U V W X

CANADA

Gander

NEWFOUNDLAND

JET STREAM

NORTH ATLANTIC OCEAN

GETTING IN LINE
Taking into account airlines' preferred routes, oceanic controllers at Gander, Newfoundland, organize aircraft approaching from different directions into position for the Atlantic crossing. This flight is entering the system on Track 3 at 35,000 feet.

EAST INTO THE NIGHT
As a result of passenger demand, time zone differences, and airport noise restrictions, North Atlantic air traffic has two peak flows: eastbound, leaving North America in the evening, and westbound, leaving Europe in the morning. Every 12 hours a new track system is prepared, to allow as many aircraft as possible to follow the most economical flight paths. Because of changing weather conditions, the track positions are rarely identical.

Graphics by
JOHN GRIMWADE

232

233

Grundy & Northedge Diagram for British Nuclear Fuels annual report showing the workings of a nuclear generator, created using Adobe Illustrator. An organizational diagram which makes strong use of bright colour and simplified graphics. By reducing the complex stages to highly stylized icons the physical structure and scale of the workings can be ignored, giving the reader a manageable understanding of the basic process.

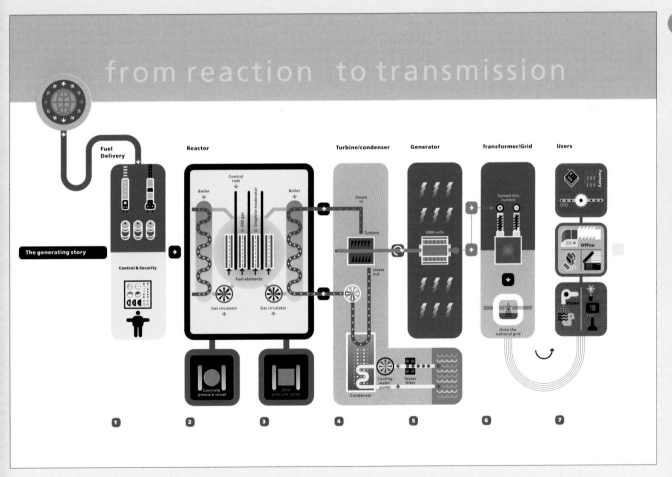

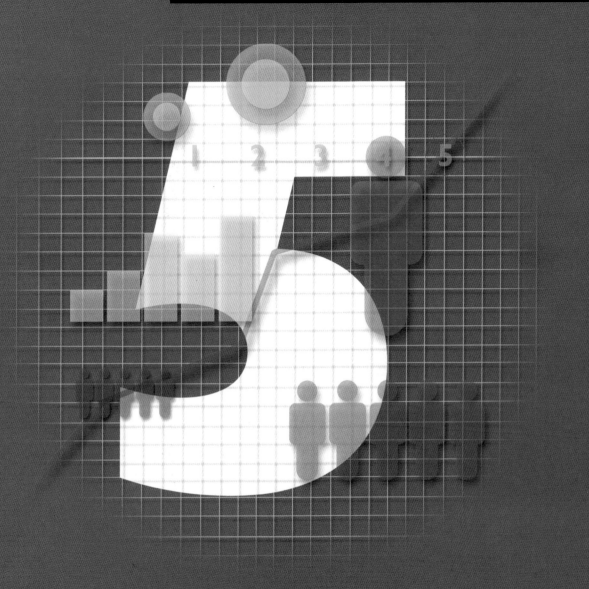

177

GLOSSARY

3-D Three dimensional, that is, an effect to give the illusion of depth on a flat page or monitor.

adaptive smoothing In some drawing and 3-D applications, smoothing which is applied to a path only where it is required, rather than uniformly.

Adobe Acrobat A proprietary 'portable document format' (PDF) file, which has fonts and pictures embedded in the document, enabling it to be viewed and printed on different computer systems.

airbrush/airbrushing A tool invented in the USA around 1900, used in illustration and photographic retouching, in which a fine spray of paint or ink is propelled by compressed air.

airbrush tool A tool used in some application programs to simulate the behaviour and effect of a mechanical airbrush.

algorithm A predetermined procedure for solving a specific problem.

algorithmically defined Usually used to describe a font in which each character is drawn 'on-the-fly' according to the calculations made by a software program, rather than residing on the computer as a predrawn font file.

aliasing The jagged appearance of bitmapped images or fonts occurring either when the resolution is insufficient or when the images have been enlarged. This is caused by the square pixels making up the image becoming visible to the eye. Sometimes called 'jaggies', 'staircasing' or 'stairstepping'.

angle of view The angle between opposite faces of a viewing pyramid (the structure created by drawing two lines from the centre of a view to its edges). The more extreme the perspective, the greater the angle.

animation The process of creating a moving image by rapidly moving from one still image to the next. Traditionally, this was achieved through a laborious process of drawing or painting each 'frame' (a single step in the animation) manually on to cellulose acetate sheets ('cels', or 'cells'). However, animations are now more commonly created with specialist software which renders sequences in a variety of formats, typically QuickTime, AVI, and animated GIFs.

anomalous data Any data which looks to be at variance with the rest of its data set.

anti-aliasing A technique of optically eliminating the jagged effect of bitmapped images reproduced on low resolution devices such as monitors. This is achieved by blending the colour at the edges of the object with its background by averaging the density of the range of pixels involved. Anti-aliasing is also sometimes employed to filter texture maps, such as those used in 3-D applications, to prevent moiré patterns.

area chart Chart that show data values as an area.

aspect ratio The ratio of the width of an image to its height, expressed as x:y. For example, the aspect ratio of an image measuring 200 × 100 pixels is 2:1.

background The area of an image upon which the principal subject or foreground sits. It may be coloured to give extra definition to the image.

bar chart/graph A chart or graph using bars to represent quantity.

baseline The line on the grid that indicates the starting point.

bas relief An image which is embossed and stands out in shallow relief from a flat background, designed to give the illusion of further depth.

Bézier curve In drawing applications, a curved line between two 'control' points. Each point is a tiny database, or 'vector', which stores information about the line, such as its thickness, colour, length and direction. Complex shapes can be applied to the curve by manipulating 'handles' which are dragged out from the control points.

binary An arithmetical system which uses 2 as its base, meaning that it can only be represented by two possible values – a 1 or a 0, on or off, something or nothing, negative or positive, small or large, etc.

binary code The computer code, based on 1 or 0, which is used to represent a character or instruction. For example, the binary code 01100010 represents a lower case 'b'.

bit A commonly used acronym for binary digit, the smallest piece of information a computer can use. A bit is expressed as one of two values – a 1 or a 0, on or off, something or nothing, negative or positive, small or large, etc. Each alphabet character requires eight bits (called a 'byte') to store it.

bit map Any text character or image composed of dots. A bit map is a map describing the location and binary state (on or off) of bits, which defines a complete collection of pixels or dots that make up an image, such as on a display monitor.

bitmapped font A font in which the characters are made up of dots, or pixels, as distinct from an outline font which is drawn from 'vectors'. Bitmapped fonts generally accompany Post-Script 'Type 1' fonts and are used to render the fonts' shape on screen (they are sometimes called 'screen' fonts). To draw the shape accurately on screen, your computer must have a bit map installed for each size (they are also called 'fixed-size' fonts), although this is not necessary if you have ATM installed, as this uses the outline, or 'printer' version of the font (the file that your printer uses in order to print it). TrueType fonts are 'outline' and thus do not require a bitmapped version.

bitmapped graphic An image made up of dots, or pixels, and usually generated by 'paint' or 'image-editing' applications, as distinct from the 'vector' images of 'object-oriented' drawing applications.

Boolean The name describing a shorthand for logical computer operations, such as those which link values – 'and', 'or' 'not', etc, called 'Boolean operators'. In 3-D applications, Boolean describes the joining or removing of one shape to or from another.

brief Written instructions detailing the parameters, scope, quality and timing of a piece of work. Given by a patron/buyer to any freelance.

bump map A bitmap image file, normally greyscale, most frequently used in 3-D applications for modifying surfaces or applying textures. The grey values in the image are assigned height values, with black representing the troughs and white the peaks. Bump maps are used in the form of digital elevation models (DEMs) for generating cartographic relief maps.

byte A single group made up of eight bits (0s and 1s) which is processed as one unit. It is possible to configure eight 0s and 1s in only 256 different permutations, thus a byte can represent any value between 0 and 255 – the maximum number of ASCII characters, one byte being required for each.

CAD (computer-aided design) Strictly speaking, any design carried out using a computer, but the term is generally used with reference to 3-D design, such as product design or architecture, where a computer software application is used to construct and develop complex structures.

camera-ready copy/art (CRC) Material, such as artwork, prepared and ready for photographic conversion to film in preparation for printing. Also called a 'mechanical', 'composite art' or 'paste-up'.

caption Strictly speaking, a headline printed above an illustration identifying its contents. Nowadays, however, a caption is generally used to mean any descriptive text which accompanies illustrative matter, and should, more accurately, be described as a 'legend'.

caricature The depiction – as a photograph, model or drawing – of a person or object with exaggerated key features or characteristics. Generally executed for comic or satirical purposes.

Cartesian coordinate system A geometry system employed in 3-D applications which uses numbers to locate a point on a plane in relation to an origin where one or more points intersect.

cartouche A decorative device, usually in the form of a scroll with rolled up ends, which is used to enclose a title or illustration.

cause-and-effect diagram A pictorial representation of the analysis of a process, identifying the cause of a problem or the solution to it, and the effect that that may have on the process. Also called a 'fishbone' diagram.

cell A single space or unit for entering data into rows and columns, such as in a spreadsheet.

clip art/clip media Collections of royalty-free photographs, illustrations, design devices and other pre-created items, such as movies, sounds and 3-D wireframes. Clip art is available in three forms – on paper, which can be cut out and pasted on to camera-ready art, on computer disc, or, increasingly, via the Web.

clipboard The place in the memory of a computer where text or picture items are temporarily stored when they are copied (or cut). The item on the clipboard is positioned in an appropriate place when 'Paste' is selected from the menu. Each process of copying or cutting deletes the previous item from the clipboard.

clipping Limiting an image or piece of art to within the bounds of a particular area.

clipping path A Bézier outline that defines which areas of an image should be considered transparent or 'clipped'. This lets you isolate the foreground object and is particularly useful when images are to be placed on top of a tint background in a page layout, for example. Clipping paths are generally created in an image-editing application such as Adobe Photoshop and are embedded into the image file when you save it in EPS format. Some page-layout applications such as QuarkXPress allow you to create clipping paths for images, but, be warned, these are only saved with the application file, so if your placed images are low-resolution FPO files which are to be replaced by the high-resolution versions, or if they are the preview files of a five-file DCS file, your clipping path will not have any effect.

clipping plane In 3-D applications, a plane beyond which an object is not visible. A view of the world has six clipping planes: top, bottom, left, right, hither and yon.

clone In image editing, to duplicate pixels in an image in order to replace defective ones, or to add to an image by duplicating parts of the same, or another, image.

column chart A vertical bar chart.

concertina bar diagram A bar diagram that compares three sets of data by adding the third set of data plots in the same plane as one of the data plots of a mirror bar chart.

constrain (1) A facility in some applications to contain one or more items within another. For example, in a page-layout application, a picture box (the 'constrained item') within another picture box (the 'constraining box'). (2) In 3-D applications, to restrict an item to a particular plane, axis or angle.

contour line A line on a map joining points of equal altitude.

control point In 'vector'-based (drawing) applications, a little knob, or point, in a 'path' (line) which enables you to adjust its shape or characteristics by means of 'Bézier control handles'.

controlling dimension One of the dimensions – either height or width – of an image, which is used as the basis for its enlargement or reduction in size.

crosshatch A line drawing style in which tonal variations are achieved by groups of parallel fine lines, each group being drawn at a different angle.

cross section An illustration of an object that shows it as if sliced through with a knife thus exposing interior details or workings.

crow quill A very fine ink pen, deriving from the days when scribes used the cut quill of a crow to write with.

cubic mapping In 3-D applications, the technique of applying one copy of a texture map (image) to each of the six sides of a cube.

cut-away An illustration of an object in which part of its outer 'shell' has been removed to show the inner workings or detail.'

cutout A halftone image in which the background has been removed to provide a freeform image. Also known as an 'outline halftone' or 'silhouette halftone'.

cylindrical (map) projection A map projection in which all lines of longitude ('meridians') run parallel with each other and at right angles to the lines of latitude ('parallels'). Cylindrical projections are highly distorted at the Polar extremities but, when wrapped around a sphere, appear normal and are thus ideal for use with 3-D applications. Cylindrical projections are widely used for navigation purposes and thus for navigational graphics. 'Gall' and 'Mercator' are typical cylindrical projections, named after the men who devised them.

cylindrical mapping In 3-D applications, the technique of applying a texture map (image) to a cylindrical 3-D object, such as applying the label on to a bottle.

data graphic Any visual representaion of data.

data point The precise value used in plotting a line graph.

degradé/degradee A graded halftone tint which, by varying the dot size, gradually changes from one edge to the other.

delineate To emphasize outlines in line drawings or artwork by making them heavier.

digital Anything operated by or created from information or signals represented by binary digits, such as a digital recording. As distinct from analog, in which information is represented by a physical variable (in a recording this may be via the grooves in a vinyl platter).

digital dot A dot generated by a digital computer or device. Digital dots are all of the same size whereas halftone dots vary, so in digitally generated halftones several dots – up to 256 (on a 16 x 16 matrix) – are required to make up each halftone dot.

digital elevation model (DEM) A collection of regularly spaced elevation data from which 3-D models of the Earth's surface can be generated. DEMs are used extensively by cartographers to produce shaded relief maps using special software.

dingbats The modern name for fonts of decorative symbols, traditionally called printer's ornaments, flowers or arabesques.

dot A term which can mean one of three things: halftone dot (the basic element of a halftone image), machine dot (the dots produced by a laser printer or imagesetter), or scan dot (strictly speaking, pixels, which comprise a scanned bitmapped image). Each is differentiated by being expressed in lpi (lines per inch) for a halftone dot, dpi (dots per inch) for a machine dot, and ppi (pixels per inch) for a scan dot – although the latter is sometimes expressed – erroneously – in dpi. Thus, because the term 'dot' is used to describe both halftone and machine dots, scan dots should always be referred to as pixels.

dot and tickle A colloquial expression for an illustration technique in which an image is composed out of small, stippled dots using a fine pen.

dot/stripe pitch The distance between the dots or pixels (actually, holes or slits in a screen mesh) on your monitor – the closer the dots, the finer the display of image.

double cranked leader line Line with two angles to reach awkward points when linking annotation/caption to part of an illustration.

double rule The term used to distinguish two lines of different thickness (in traditional typesetting, made from brass) from two lines of the same thickness called a 'parallel rule'.

draw(ing) application Drawing applications can be defined as those which are object-oriented (they use 'vectors' to mathematically define lines and shapes), as distinct from painting applications which use pixels to make images ('bitmapped graphics') – some applications do combine both.

drop shadow A shadow projected on to the surface behind an image or character, designed to lift the image or character off the surface.

elevation The drawn vertical projection of an object, typically front, side, or end.

ellipse An oval shape which corresponds to the oblique view of a circular plane.

embossing (1) Relief printing or stamping in which dies are used to impress a design into the surface of paper, cloth or leather so the letters or images are raised above the surface. (2) Simulation of this effect on a computer.

end cap In drawing applications, the method in which the end of a line is rendered; or, of an object; in 3-D applications, the closing of an open end left by a lathe, sweep or extrude operation and time.

engraving (1) A block or plate made from wood or metal into which a design or lettering has been cut, engraved or etched. (2) A print taken from an engraved print or block. (3) Computer simulation of this effect.

error diffusion In digital scanning, the enhancement of an image by averaging the difference between adjacent pixels. In graphics applications this technique is more commonly referred to as 'anti-aliasing' which, in turn, uses a technique known as 'interpolation'.

exploded view An illustration of an object displaying its component parts separately – as though it were exploded – but arranged in such a way as to indicate their relationship within the whole object when assembled.

extrude The process of duplicating the cross-section of a 2-D object, placing it in a 3-D space at a distance from the original and creating a surface which joins the two together. For example, two circles that become a tube.

eye point In a 3-D world, the position of the viewer relative to the view.

fall off In a 3-D environment, the degree to which light loses intensity away from its source.

feather(ing) The gradual fading away of the edge of an image or part of an image to blend with the background. Feathering tools are a feature of image-editing and painting applications.

fill In graphics applications, the contents, such as colour, tone or pattern, applied to the inside of a closed path or shape, including type characters.

fill characters The characters inserted between specified tab stops in some applications.

fine rule A rule, traditionally of hairline thickness. In graphics applications this can mean a rule of any thickness between 0.20–0.50 points, but which is invariably defined as 0.25 points. (see *also* **hairline rule**)

flow chart A diagrammatic representation of a process.

free form shape A shape that is not regular or geometric.

fresnel factor In a 3-D environment, the brightening of the edge of an object by increasing the intensity of reflection along the edge, giving a more realistic visual effect.

ghosting An illustration technique in which the outer layers of an object are faded to reveal its inner parts, which would not normally be visible. The technique is used particularly in technical illustration, to show parts of an engine covered by its casing.

Gouraud shading In 3-D applications, a method of rendering by manipulating colours and shades selectively along the lines of certain vertices, which are then averaged across each polygon face in order to create a realistic light and shade effect.

graphic device Any illustration or drawn design.

graph paper Specialist paper marked with a series of very fine and thicker lines for the purposes of precise plotting and drawing of graphs.

graticule A grid system of lines used to provide reference points on an image, particularly the lines of latitude ('parallels') and longitude ('meridians') on a map.

grid In some applications, a user-definable background pattern of equidistant vertical and horizontal lines to which elements such as guides, rules, type, boxes, etc., can be locked, or 'snapped', thus providing greater positional accuracy.

ground plane In some 3-D applications, a reference plane oriented on the *x–z* axis, providing a reference for the ground.

guides In many applications, nonprinting horizontal and vertical lines which can be placed at inconsistent intervals and which help you to position items with greater accuracy.

hachures Parallel lines used in hill-shading on maps, their closeness indicating the steepness of gradient.

hairline rule Traditionally, the thinnest line that it is possible to print. In applications which provide it as a size option for rules, it is usually 0.25 points wide.

high-profile Any item which stands apart or visibly proud of its surroundings.

high resolution monitor Monitor capable of displaying 24-bit colour.

histogram A graphic representation of data, usually in the form of solid vertical or horizontal bars. Some image-editing applications use histograms to graph the number of pixels at each brightness level in a digital image, thus giving a rapid idea of the tonal range of the image so that you can tell if there is enough detail to make corrections.

icon In a computer 'graphical user interface', a graphic representation of an object, such as a disc, file or tool, or of a concept or message.

ideogram A symbol used to indicate a concept or emotion, such as joy, sadness, warmth, etc.

illuminate/illuminated The technique of embellishing letters, pages, manuscripts, etc., by using gold, silver and colours, common in medieval times.

inked art Camera-ready art which is prepared firstly in pencil and then completed in ink.

intersection In some drawing and page-layout applications, a feature that allows you to create a shape from two or more others that overlap or intersect each other.

intrinsic mapping In 3-D applications, a method of mapping a texture to a surface using the object's own geometry as a guide to placement rather than the geometry of another shape such as a cylinder or sphere.

isometric Where the plane of projection is at equal angles to the three principal axes of the object shown.

isotype A method for graphically presenting statistical information in which pictograms or ideograms are used to convey meaning, a technique pioneered by Otto and Marie Neurath's Isotype Institute.

key An explanatory list of symbols or colour coding used on a map, graph, chart or any representation of information in visual form.

keyline A line drawing indicating the size and position of an illustration in a layout.

keyline view In some applications, a facility which provides an outline view of an object or illustration without showing attributes such as colours and line thicknesses.

leader(s) A row of dots, dashes or other characters, used to guide the eye across a space to other relevant matter. In some applications, leaders can be specified as a 'fill' between tab stops in text.

leader line/leader rule A line keying elements of an image to an annotation or a caption.

linear motion style In 3-D applications, a calculation of the unknown values between two known keyframe values of an animation based on calculating the shortest distance between the two.

line tool In graphics applications, the tool used to draw lines and rules. If the tool can only draw horizontal and vertical rules it is usually called an 'orthogonal' line tool.

logarithm A figure representing the power to which a fixed number or base must be raised to produce a given number. Used to simplify calculations as the addition and subtraction of logarithms is equivalent to multiplication and division.

logarithmic scale A scale based on logarithms. These never start from zero, and can be in any base but base ten is the most frequently used.

matrix A rectangular array of elements in rows and columns that is treated as a single entity.

median The middle value of a series of values arranged in order of size.

midtones/middletones The range of tonal values in an image anywhere between the darkest and lightest, usually referring to those approximately halfway.

mirror bar chart A bar chart that compares two sets of data where they are placed either side of a baseline.

mitred/mitre The bevelled ends of right-angled frame rules used in traditional typesetting. Also used in software applications as a method of defining the 'end caps' in rules or lines.

mono(chrome)/monochromatic An image of varying tones reproduced in a single colour.

multiple graphs Graphs which show the plots of several sets of data, used for comparison of similar data.

negative value Values less than zero, to be subtracted from others or from zero.

normal(s) In 3-D objects, the direction that is perpendicular to the surface of the polygon to which it relates.

oblique projections Projections drawn with one plane of the object drawn face on, and the receding planes angled to one side. Although any angle other than 90° can be used 45°, 30°, and 60° are the most common.

one-point perspective Projection with one vanishing point.

orthographic (projection) An illustration technique in which there is no perspective (or the perspective is infinite), thus giving parallel projection lines.

outline font A vector-based digital font drawn from an outline which can be scaled or rotated to any size or resolution, as distinct from 'bitmapped fonts' comprised of pixels and used primarily for screen display in the case of PostScript fonts. Also called 'printer fonts' or 'laser fonts' (because they are essential for rendering fonts accurately when output on laser printers and imagesetters), or 'scaleable fonts'.

Oxford corners A box rule where the lines project beyond the frame corners.

Oxford rule The combination of a thick and a thin rule.

paint(ing) applications Applications which use bitmaps of pixels to create images rather than the 'vectors' that describe lines in drawing applications (called 'object-oriented'), although some applications combine both.

PANTONE® The proprietary trademark for Pantone Inc.'s system of colour standards, control and quality requirements, in which each colour bears a description of its formulation (in percentages) for subsequent printing. The PANTONE MATCHING SYSTEM® is used throughout the world so that colours specified by any designer can be matched exactly by any printer.

parallax The apparent movement of two objects relative to each other when viewed from different positions.

parallel rule The term used to distinguish a rule comprising two lines of the same thickness (made from brass in traditional typesetting) from one comprising two lines of different thickness called a 'double rule'.

Pareto diagram A diagram which illustrates the theory (the 'Pareto principle') that 80 percent of the effects of a situation come from only 20 percent of possible causes.

peaking A method of digitally sharpening images by using a filter that increases the difference in density where two tonal areas meet. Also called 'sharpen edges'.

perspective A technique of rendering 3-D objects on a 2-D plane, duplicating the 'real world' view by giving the same impression of the object's relative position and size when viewed from a particular point – the shorter the distance, the wider the perspective; the greater the distance, the narrower the perspective.

phong shading A superior but time-consuming method of rendering 3-D images, which computes the shading of every pixel. Usually used for final 32-bit renders.

pictogram/pictograph A simplified, pictorial symbol distilled to its salient features to represent an object or concept.

pie chart An illustration which shows the divisions of a whole quantity as segments of a circle or ellipse.

pitch In 3-D construction, the rotation around the x-axis.

pixel Acronym for picture element. The smallest component of a digitally generated image, such as a single dot of light on a computer monitor. In its simplest form, one pixel corresponds to a single bit; $0 = $ off, or white, and $1 = $ on, or black. In colour or greyscale images or monitors, one pixel may correspond to several bits: an 8-bit pixel, for example, can be displayed in any of 256 colours (the total number of different configurations that can be achieved by eight 0s and 1s).

pixelation/pixellization An image that has been broken up into square blocks resembling pixels, giving it a 'digitized' look.

plot area The area of a graph defined by the data. The area containing the data points.

point In object-oriented drawing applications, the connections ('Bézier points') which mathematically define the characteristics of line segments, such as where they start and end, how thick they are, and so on (each point is a 'vector' or tiny database of information). Lines are manipulated by dragging 'control handles' (sometimes called 'guidepoints') from the point, which act on the line like magnets.

point markers In line graphs these are used to mark the position of the precise data values used to plot the line.

polygon The smallest unit of geometry in 3-D applications, the edges of which define a portion of a surface.

polygon resolution The detail in a 3-D scene, defined by the number of polygons in a surface, which, in turn, determines the detail of the final render – the more polygons, the finer the detail.

polygon tool A tool in some applications with which you can create irregular-shaped boxes, in which text, pictures or fills can be placed.

polyline In 3-D applications, a line with more than two points which defines a sequence of straight lines.

polymesh A 3-D object comprising shared vertices in a rectangular shape.

primitive The basic geometric element from which a complex object is built.

projection mapping Texture mapping on to a 3-D object where the texture appears to project through the image surface.

puck The rather more complex 'mouse' used in CAD systems.

radial fill A feature of some applications in which an item can be filled with a pattern of concentric circles of graduated tints.

raster image An image defined as rows of pixels or dots.

raster(ization) Deriving from the Latin word 'rastrum', meaning 'rake', the method of display (and of creating images) employed by video screens, and thus computer monitors, in which the screen image is made up of a pattern of several hundred parallel lines created by an electron beam 'raking' the screen from top to bottom at a speed of about one-sixtieth of a second. An image is created by the varying intensity of brightness of the beam at successive points along the raster. The speed at which a complete screen image, or frame, is created is called the 'frame' or 'refresh' rate.'

rate determining factor (RDF) The rate at which progress is determined, defined as the slowest part of any procedure or process. For example, the rate at which you may be able to complete a design job will depend on how long it takes the slowest contributor (an author or illustrator, for example) to complete their task, in which case they are the rate determining factor. It could even be you.

raytracing Rendering algorithm which simulates the physical and optical properties of light rays as they reflect off a 3-D model, producing realistic shadows and reflections.

readable The ease and comfort with which text can be read.

reflection tool In some applications, a tool for transforming an item into its mirror image, or for making a mirror-image copy of an item.

relational diagram A diagram showing the relative association of objects in the physical world, for example maps or plans.

relief map A cartographic map in which elevation is rendered to simulate three dimensions.

render(ing) The process of creating a 2-D image from 3-D geometry to which lighting effects and surface textures have been applied.

rescale Amending the size of an image by proportionally reducing its height and width.

reverse image An image that has been reversed, either horizontally or vertically, so that what was on the left of the image is now on the right, or what was at the top is now at the bottom, as in a mirror image.

reverse out/reverse type To reverse the tones of an image or type so that it appears white (or another colour) in a black or coloured background. Also called 'dropout', 'save out' or 'knockout'.

roll In a 3-D environment, the rotation of an object or camera around the z-axis.

rotation tool In some graphics applications, a tool which, when selected, enables you to rotate an item around a fixed point.

roughness map In some 3-D applications, the use of a texture map to control surface roughness.

rule A printed line. The term derives from the Latin 'regula' meaning 'straight stick' and was used to describe the metal strips of type height, in various widths, lengths and styles, which were used by traditional typesetters for printing lines.

ruler In some applications, the calibrated ruler at the edges of a document window, in a preferred unit of measure.

ruler guide In some applications, a nonprinting guide which you position by clicking on the ruler and dragging it to the desired location.

same size (S/S) An instruction that original artwork/illustrations should be reproduced without enlargement or reduction.

sample size The number of pixels, or the amount of data, used as a sample to assess information about an image or about other digital files, such as sounds.

scale an animation To lengthen or shorten the length of time of an animation.

scale/scaling The process of working out the degree of enlargement or reduction required to bring an image to its correct reproduction size.

scale drawing A drawing such as a map or plan which represents the subject matter in proportion to its actual size and to a specified scale. For example an object drawn to a scale of 1:10 means that one unit of measurement on the drawing is equivalent to 10 of the same units on the object at full size – so one inch on the drawing equals 10 inches at full size.

scan dot The resolution of a scanning device, measured – like machine dots – in dots per inch. However, the formula for calculating scan dot resolution does not relate to that of calculating machine dot resolution. A rough rule of thumb is that for images that will eventually be printed with a halftone screen ruling greater than 133lpi, the scan resolution should be 1.5 times the lpi screen ruling, and for screens equal to, or less than, 133lpi, the scan resolution should be 2.0 times the lpi ruling. Scanning at higher resolutions will not make any difference because the halftone dots will not be small enough to reproduce the extra detail.

scatter diagram/graph A graph used to analyse the cause and effect relationship between two variables where two sets of data are plotted on x and y axes, the result being a scattering of unconnected dots.

score In some 3-D applications, the term describing time information about each object and property in a scene.

sepia A brown colour and also a monochrome print in which the normal shades of grey appear as shades of brown.

shading In 3-D applications, the resulting colour of a surface due to light striking it at an angle.

shadow (area) The areas of an image which are darkest or densest.

sharpen(ing) Enhancing the apparent sharpness of an image by increasing the contrast between adjacent pixels.

skew(ing) A feature built into some graphics applications, allowing type characters or images to be slanted. Also known as 'shearing'.

185

skin In 3-D applications, a surface stretched over a series of 'ribs', such as an aircraft wing.

smoothing The refinement of bitmapped images and text by a technique called 'anti-aliasing' (adding pixels of an 'in-between' tone). Smoothing is also used in some drawing and 3-D applications, where it is applied to a path to smooth a line, or to 'polygons' to tweak resolution in the final render.

smooth point A Bézier control point which connects two curved lines, forming a continuous curve.

snap to In many applications, a facility which automatically guides the positioning of items along invisible grid lines which act like magnets, aiding design and layout.

specular map In 3-D applications, a texture map – such as those created by noise filters – which is used instead of specular colour to control highlights.

specular reflectance The reflection (as by a mirror) of light rays at an angle equal to the angle at which it strikes a surface (angle of incidence).

spherical map(ping) In 3-D applications, a technique of mapping a rectangular image to a sphere – the rectangle first becomes a cylinder, which is then wrapped around a sphere by pinching the top and bottom of the cylinder into single points, or 'poles'.

spline The digital representation of a curved line that is defined by three or more control points, common in 3-D applications.

split bar chart A bar chart where the individual components in a bar are broken down into its individual parts and indicated. The component bars are placed end to end, either as a vertical column chart or horizontal bar graph. Also known as a stacked bar.

spotlight In 3-D applications, a beam of light whose beam is shone as a cone.

spreadsheet An application that allows you to make complex calculations encompassing almost any user defined criteria. A spreadsheet employs a table of rows and columns, and the spaces in the grid, known as cells, can be moved and mathematically manipulated. Some spreaadheet applications will also generate graphics (3-D in some cases) from the data entered into the cells.

square corner tool A feature of most graphics applications, this is a tool used for drawing squares and rectangles. Also called a 'rectangle tool'.

stacking order The position of items relative to others, in front or behind, in software.

starburst In some applications, the shape that the pointer assumes when certain transformation tools are selected.

stylize/stylization The imposition of a graphic style on an image, as distinct from a realistic representation of it.

surface In 3-D applications, the matrix of control points and line end points underlying a mapped texture or colour.

surface geometry In 3-D applications, the geometry that underlies a surface, becoming visible when the surface is simplified.

symbol A figure, sign or letter that represents an object, process or activity. A computer icon, for example, is a pictorial symbol.

symmetrical point A Bézier control point that connects two curved lines, forming a continuous curve. Also called a 'curve point'.

table A set of facts or figures displayed systematically, particularly in columns.

tangent line In a 3-D environment, a line passing through a control point of a spline at a tangent to the curve. The tangent line is used to adjust the curve.

taper Referring to graduated tones and colours, the progression of one tone or colour to the next.

taper angle The direction in which graduated tones or colours merge into one another.

template (1) A shape used as a drawing aid. (2) A document created with prepositioned text and images, used as a basis for repeatedly creating other similar documents.

texture mapping/surface mapping In 3-D environments, the technique of wrapping a 2-D image around a 3-D object.

three-point perspective A projection with three vanishing points.

tile/tiling The repetition of a graphic item, with the repetitions placed side-by-side in all directions so that they form a pattern – just like tiles.

tilt In a 3-D environment, a camera that performs vertical pans (up and down) about its horizontal axis.

topographic structure A detailed description of the natural and artificial features of a terrain.

tracking The adjustment of space between characters in a selected piece of text. As distinct from 'kerning', which only involves pairs of characters.

transformation tool In some applications, the name given to tools that change the location or appearance of an item, such as 'scale' or 'reflection'.

transposing (1) Exchanging the position of any two items of text, or two images. (2) In a bar chart shifting the alignment of the bars.

trend line In a scatter graph, a line drawn through a concentration of data points to pick it out.

trim curve In a 3-D environment, a curve on the surface of one object where it is intersected by another, allowing you to trim away parts of the surface.

tritone A halftone image that is printed using three colours. Typically, a black-and-white image is enhanced by the addition of two colours, for example process yellow and magenta when added to black will produce a sepia-coloured image.

uniform smoothing In some 3-D applications, uniform smoothing converts the surface of a model into a grid of evenly spaced polygons.

union In drawing applications, the combining of two or more shapes into one.

up vector In a 3-D environment, a line perpendicular to the view point of a camera that allows the camera object to be rolled around the view point.

u-v coordinates In a 3-D environment, a system of rectangular 2-D coordinates used to apply a texture map to a 3-D surface.

vector Information giving both magnitude and direction.

vector drawing program A program that uses mathematical points to define lines and shapes.

velocity In 3-D animation, the rate of change in an object's location relative to time.

vertex In a 3-D environment the x, y and z ocations at each corner of a polygon or control point.

vertex animation In 3-D animation, varying the shape of an object by animating its surface control points.

vertical alignment The placement of items such as images or lines of text in relation to the top and bottom of a page, column or box.

vertical centring The equidistant positioning of text ('vertical justification') or any other item from the top and bottom of a page, column or box.

volume chart Used for comparing data vastly different in value, and for representing large data. This is because it uses three dimensions as opposed to two and so can representvalues up to ten times the size of those represented by an area chart without occupying more space.

x, y coordinates The point at which data is located on two-dimensional axes: horizontal (x) and vertical (y) – for example, on a graph, monitor, bitmap, etc. Three-dimensional coordinates are known as 'x, y, z coordinates' (or axes).

yaw In a 3-D environment, rotation around the y-axis.

z-buffer render A 3-D renderer which solves the problem of rendering two pixels in the same place (one in front of the other) by calculating and storing the distance of each pixel from the camera (the 'z-distance'), then rendering the nearest pixel last.

zero base The horizontal line which cuts across the y-axis at value zero. Any value above the zero base is thus a positive value, and any value below the zero base a negative one.

INDEX

188

ACKNOWLEDGEMENTS

We are grateful to the following for their kind help and for allowing their work to be included in this volume.

John Grimwade for 'Round the World Balloons' and 'How Hurricanes Happen' from *Popular Science*, 'Pipe Organ' from the *Smithsonian Magazine*, 'New York Grand Central Station', 'Angkor Wat', 'Map of Florence', 'The Yangtze River' and 'North Atlantic Air Traffic' from *Condé Nast Traveler*; Grundy & Northedge for 'Heading Icons' from *Bloomberg Magazine*, 'Icons' from the *Camelot Annual Report*, 'Five Financial Crashes' from *Worldlink Magazine*, 'Movie-goer's View of San Francisco' from *The Sunday Times*, and 'How Nuclear Generators Work' from the *British Nuclear Fuels Annual Report*; Nigel Holmes for 'Diagram of Breast Examination' from *Self Magazine*, 'Greg Louganis' Reverse Pike Dive' from *Olympic Dreams* (Rizzoli), 'Map of the Brain' and 'The Bombing of the Uffizi Gallery' from *Time Magazine*, 'Gas Engine' from *Attaché Magazine*, and 'The National Debt' and 'US Federal Budget' from *Understanding*; Bill Le Bihan for the 'Four Stroke Engine Cycle'; Annabel Milne for 'Teeth cleaning' and 'Angiogenesis' from the *Antisoma Prospectus*; and Rowson Holbrook for 'Map of a Railway Track Modification for the North London Line'.